Environmental Sustainability and Risk Management Strategies

Dr. Gummadi Venkata Rao, Professor

Mr. Katukuri Hajarath, Assistant Professor

INDIA • SINGAPORE • MALAYSIA

ISBN
Paperback 979-8-89984-148-4
Hardcase 979-8-89984-149-1

Contents

Chapter-1

Reduce, Reuse, and Recycle

1.1. Three R's - Reduce, Reuse, and Recycle

The three R's represent "Reduce, Reuse, and Recycle." This environmental trio serves as principles that encourage us to protect our planet and utilize resources judiciously.

1. Reduce

"Reducing" refers to the practice of minimizing consumption, such as opting for fewer plastic bottles or selecting items with minimal packaging. It focuses on decreasing waste from the outset. To "reduce" means to avoid excessive purchases initially.

This concept emphasizes the importance of using resources judiciously to prevent them from ultimately contributing to landfill waste. For example, using your own reusable coffee cup or water bottle exemplifies the act of "reducing." By adopting this habit, you contribute to lowering the demand for single-use containers and help decrease overall waste.

How Long Does Plastic Decompose?

Plastic takes a long time to decompose. For some materials, it can take tens or even hundreds of years. The main reason is because plastics do not easily break down by natural processes. But nature can break down things like food or leaves in just months. From grocery bags to plastic straws, let us examine how long it takes for plastic to decompose.

Cigarette Filter (10 years)

Cigarette filters are typically composed of a plastic known as cellulose acetate. This material is both robust and resistant to degradation, often requiring a lengthy period of 10 to 15 years to fully decompose. While they do break down faster than certain other plastics, their durability allows cigarette filters to remain in the environment for extended periods, posing a significant threat to nature.

Plastic Grocery Bag (20 years)

Biodegradation of Plastic Grocery Bags Plastic grocery bags are primarily composed of polyethylene, which comes in two varieties: high-density polyethylene (HDPE) and low-density polyethylene (LDPE). HDPE bags are characterized by their thinner and sturdier design, whereas LDPE bags offer greater flexibility and lower density. Both forms of plastic can persist in the environment for more than two decades.

Coffee Cups (50 years)

Biodegradation of Coffee Cups Plastic coffee cups, like various other plastic products, require an extensive period to decompose. It can take up to fifty years for these cups to break down completely. These cups are primarily constructed from polystyrene, a type of plastic that resists natural degradation processes. Consequently, they remain in the environment for an exceedingly long time, which poses a significant threat to ecological health.

Polyester Shirts (50+ years)

Biodegradation of Polyester Shirts Polyester shirts also have a prolonged decomposition period. Like coffee cups, these garments can persist in the environment for several decades. Typically made from synthetic fibres engineered for durability and resistance to moisture and bacteria, polyester remains intact for an extended duration, allowing these shirts to linger in the environment.

Aluminium Can (200 years)

Aluminium is fundamentally different from plastic. It requires more than 200 years to decompose due to its metallic nature. Although aluminium can corrode or rust over time, this process occurs at a very gradual pace. The metal's resistance to moisture and oxygen, which typically contribute to material degradation, allows aluminium cans to persist in the environment for several decades before they eventually break down.

Drinking Straws (200+ years)

Recently, there has been a growing movement to reduce the use of plastic straws, with an emphasis on alternatives such as paper or reusable options. Plastic straws contribute to ocean pollution and pose a danger to marine life, as they can be ingested by animals. The lengthy decomposition period of over 200 years for drinking straws presents a significant environmental concern.

Disposable Diapers (400+ years)

The decomposition of disposable diapers is a lengthy process, primarily due to their construction from various synthetic materials, including plastics and super-absorbent polymers. These components are engineered for durability and moisture resistance, which are essential for the diaper's functionality. However, this very durability hinders their breakdown in the environment. As a result, these diapers can persist for centuries, contributing to significant long-term environmental challenges in landfills.

Plastic bottles also exhibit a slow decomposition rate, as they are typically manufactured from a type of plastic known as PET (polyethylene terephthalate). The resilience of PET plastic against environmental degradation is a key factor in the environmental issues associated with plastic bottles. Conversely, PET bottles rank among the most widely recycled plastics. Recycling these bottles presents an environmentally friendly solution to mitigate plastic waste by converting them into new products.

Single-use plastics, such as plastic bottles, disposable diapers, and drinking straws, pose the greatest threat to the environment and marine ecosystems. Many individuals discard these items after a single use, and they can remain in the environment for hundreds of years. We advocate for a reduction in plastic usage and increased recycling efforts to address the challenges posed by plastic waste.

2. Reuse

The term "reuse" refers to the practice of utilizing items again rather than discarding them. There are numerous innovative methods to repurpose various items found in your home. For instance, old cardboard boxes can be transformed into engaging projects or utilized as storage solutions Image of Reduce, Reuse and Recycle as shown in figure 1. This approach not only conserves financial resources but also minimizes the number of boxes that are disposed of.

Figure 1 Image of Reduce, Reuse and Recycle

Another illustration of reuse is employing an empty glass jar for storing leftovers instead of discarding it. This practice extends the jar's utility and decreases the demand for new storage containers.

3. Recycle

"Recycling" involves the collection and processing of materials such as paper, plastic, and glass, enabling them to be converted into new products. This process provides a second life to old items, preventing them from being sent to landfills. In the United States, approximately 69 million tons of materials are recycled annually.

Paper and cardboard account for about two-thirds of all recycled materials, while metals represent around 13%. Glass, plastic, and wood collectively make up approximately 4 to 5%. Food waste remains a critical issue, with 30-40% of all food produced in the United States being discarded. This is a significant concern as it results in the wastage of edible food and contributes to environmental challenges.

1.2 Recycling process

Recycling: Definition Recycling refers to the method of gathering and processing discarded materials to transform them into reusable resources. It plays a crucial role in reducing waste in contemporary society. A material is considered recycled when it is utilized, repurposed, or recovered. This practice helps prevent or lessen the loss of valuable resources and decreases the demand for new raw materials, thereby reducing energy consumption. Additionally, recycling helps to control or diminish environmental pollution, such as air and water contamination, which can occur from incineration and landfilling processes. Recyclable materials encompass a variety of items, including paper (such as newspapers, cardboard, and mixed paper), glass (in colors like amber, green, and flint), cans (including aluminum, ferrous, and bimetal), and various types of plastics (such as PET, HDPE, PS, PVC, PP, and LDPE), among others.

1.2.1. Types of Recycling

Recycling can be classified into three types. They are primary, secondary, and tertiary recycling. Tertiary recycling is further sub classified into internal and external recycling. Figure 1 shows the schematic view of the types of recycling.

a. **Primary Recycling**

\Primary recycling refers to the process of reusing and recovering materials in their original form without any modifications after the recycling process. It can be characterized as the practice of second-hand use or reuse, where the intended purpose of the material remains unchanged. Examples include donating items to friends, family members, or charitable organizations, as well as selling them.

b. **Secondary Recycling**

Secondary recycling entails making significant alterations to materials or products without employing chemical processes. This can involve activities such as cutting and reshaping various items for arts and crafts or transforming envelopes into smaller pieces for use as scrap paper.

c. **Tertiary Recycling**

Tertiary recycling refers to the reprocessing of materials or products through chemical methods or heat. Examples include the melting of metals, the chemical treatment of used paper, and the crushing of plastic bottles to create new items.

d. **External Recycling**

Tertiary recycling is classified as external recycling when the recovery and reprocessing of materials or products involve public participation. This includes activities such as sorting waste and placing it in recycling bins for collection and transport to reprocessing facilities.

e. **Internal Recycling**

Tertiary recycling is considered internal recycling when the recovery of materials or products occurs without public involvement, specifically within factories and manufacturing plants.

1.2.2. Steps involved in Recycling process

Recycling process involves the following steps that assists in the minimization of waste material or products thereby reducing their environmental impacts. The steps involved in the recycling process is shown in Figure 2,

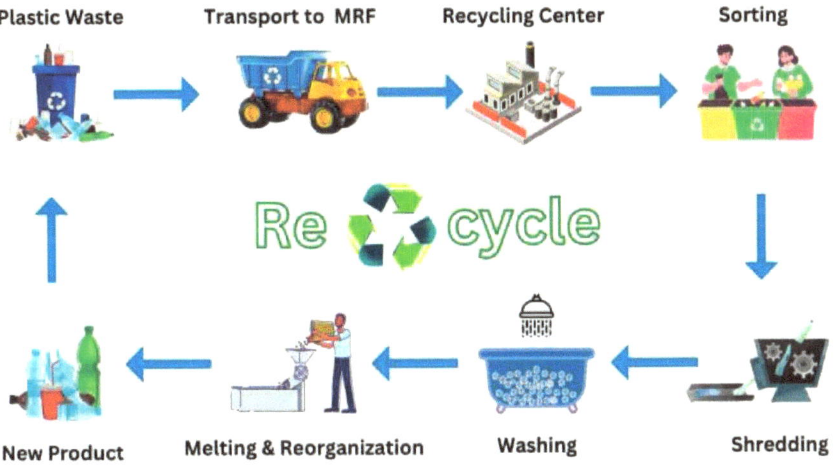

Figure 2 Steps in Steps involved in Recycling

a) Collection

The Recyclable materials from municipal solid waste is collected by the following methods.

1. Collection of segregated or source-separated recyclables. They are collected by the waste collectors with and without subsequent processing.
2. The materials for recycling can also be collected from the transfer stations or any other centralized recovery facility.

If the waste is a mixed municipal solid waste, then they must be segregated and collected in different coloured bins. Figure 3 depicts the colour coded bins for segregated collection of waste. Green bins or sacks are for green waste e.g. Food and garden waste. Blue colour bins are for recyclables and black is for non-recyclables. Once segregated, the recyclables are delivered to waste collectors for further processing.

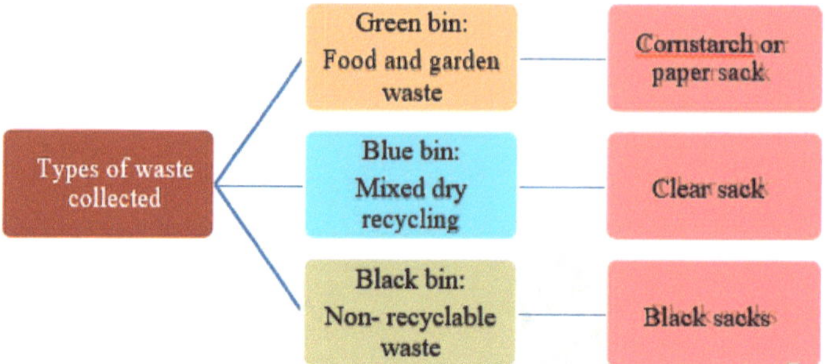

Figure 3 Collection of different waste in bin & sacks

b) Mixed Dry Recycling

Clear bags are designated for the collection of this waste type; it is important not to combine them with black bag waste and to regularly flatten plastic bottles. Below is a list of materials suitable for mixed dry recycling that can be placed in the blue bin: i. Cans, which include: food tins, beverage cans, empty aerosol containers, tinfoil, and foil containers from ready meals (after food has been removed). ii. Cardboard, which encompasses: cardboard boxes, food packaging made of cardboard, greeting cards, inserts from toilet rolls and kitchen rolls, and food and drink cartons (with liquids emptied). iii. Paper, which consists of: newspapers, magazines, catalogues, telephone directories, paperback books, envelopes, junk mail, and shredded paper. iv. Glass, which includes: glass bottles for drinks, glass jars, and other glass containers. v. Plastic, which comprises: plastic bottles, yogurt pots, containers for ready meals, ice cream tubs, and supermarket carrier bags. 3.1.2 Non-Recyclable Waste

The collection of non-recyclable materials varies by property type, occurring either from the black bin, black sacks, or large communal wheeled bins. Non-recyclable items include disposable diapers, sanitary products, incontinence pads, polystyrene packaging, pet waste or cat litter, broken dishes, cigarette butts and ashtray contents, as well as coal and wood ash.

c) Categories of waste collection

There are three categories of recycling material collection centres as shown in Figure 4.

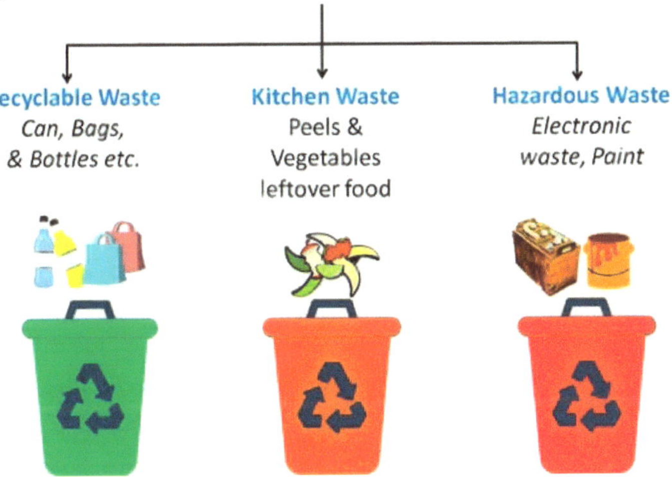

Figure 4 Categories of waste collection

Drop-off Centres: Drop-off centres require waste producers to transport recyclable materials to a designated location, which may be a fixed or mobile collection station or the recycling facility itself. While these centres are the easiest to implement, they often experience low and inconsistent volumes of materials.

Buy-back Centres: In contrast, buy-back centres operate by purchasing cleaned recyclables, thereby offering a clear incentive for participation, and ensuring a more reliable supply. The processed materials can subsequently be resold, ideally generating profit. However, government subsidies are essential for the sustainability of buy-back centres, as reported by the U.S. National Waste & Recycling Association, which states that the average cost to process a ton of material is US$50, while it can only be resold for US$30.

Curb side Collection: Curb side collection encompasses various systems that primarily differ in the stage at which recyclables are sorted and cleaned. The main types include mixed waste collection, commingled recyclables, and source separation. Typically, a waste collection vehicle is responsible for picking up the waste. An illustration of a truck designated for collecting recyclable materials is depicted in Figure 5.

Figure 5 Collecting recyclable materials

d) Sorting

Once commingled recyclates are collected and delivered to a central collection facility, the different types of materials must be sorted. This is done in a series of stages, many of which involve automated processes such that a truckload of material can be fully sorted in less than an hour. Some plants can now sort the materials automatically, known as single-stream recycling. In plants, a variety of materials are sorted such as paper, different types of plastics, glass, metals, food scraps and most types of batteries. A 30 percent increase in recycling rates has been seen in the areas where these plants exist. Sorting can be done by various methods. They are

i. By hand

ii. By Automated machinery

iii. By strong metals

By hand

The commingled recyclable material is removed from the collection vehicle and placed on a conveyor belt spread out in a single layer. Large pieces of corrugated fibre board and plastic bags are removed by hand at this stage, as they can cause later machinery to jam.

By Automated machinery

Automated machinery such as disk screens and air classifiers separates the recyclable material by weight, splitting lighter paper and plastic from heavier glass and metal. Cardboard is removed from the mixed paper and the most common types of plastics are collected. This separation is usually done by hand but has become automated in some sorting centres: a spectroscopic scanner is used to differentiate between different types of paper and plastic based on the absorbed wavelengths, and subsequently divert each material into the proper collection channel.

By strong metals

Strong magnets are used to separate out ferrous metals, such as iron, steel and tin cans. Non-ferrous metals are ejected by magnetic eddy currents in which a rotating magnetic field induces an electric current around the aluminium cans, which in turn creates a magnetic eddy current inside the cans. This magnetic eddy current is repulsed by a large magnetic field, and the cans are ejected from the rest of the recyclate stream.

Finally, glass is sorted based on its colour: brown, amber, green, or clear. It can either be sorted by hand, or via an automated machine that uses coloured filters to detect different colors. Glass fragments smaller than 10 millimetres (0.39 in) across cannot be sorted automatically, and are mixed together as "glass fines.

a) Rinsing

   Food packaging must be free of any organic materials. Any organic waste should be disposed of in a biodegradable waste bin or buried in a garden. To prevent the presence of organic matter, it is essential to rinse the packaging materials before placing them in a trash bag.

b) Processing & Recycling

   While numerous government initiatives focus on home recycling, it is important to note that 64% of waste originates from industrial sources. Many recycling programs within the industry prioritize the cost-effectiveness of recycling efforts. Cardboard packaging, due to its widespread use, is frequently recycled by companies involved in the distribution of packaged goods, such as retail outlets, warehouses, and distributors. Other industries may focus on niche or specialized products, depending on the specific waste materials they handle. Manufacturers of glass, lumber, wood pulp, and paper typically work with commonly recycled materials. Additionally, independent tire dealers may collect and recycle old rubber tires for profit. Table 1 illustrates the various methods of waste processing and recycling. Recycling rates for metals tend to be relatively low. Although metals are inherently recyclable, the existing metal stocks in society represent significant above-ground resources. However, the recycling rates for many metals remain disappointingly low. Particularly concerning are the recycling rates of rare metals utilized in devices such as mobile phones, hybrid car battery packs, and fuel cells. Without a substantial increase in recycling rates for these critical metals in the future, their availability for modern technological applications may be jeopardized.

The military engages in the recycling of certain metals. The U.S. Navy's Ship Disposal Program employs ship breaking techniques to recover steel from decommissioned vessels. Additionally, some ships are intentionally sunk to form artificial reefs. Uranium, a highly dense metal, possesses advantages over lead and titanium for various military and industrial applications. The residual uranium from its conversion into nuclear weapons and reactor fuel is referred to as depleted uranium, which is utilized by all military branches for armor-piercing projectiles and protective shielding. In the construction sector, concrete and old road surface materials are often recycled, with companies profiting from the sale of these waste products. Certain industries, particularly the renewable energy sector and solar photovoltaic technology, are proactively establishing recycling initiatives even before their waste production becomes significant, anticipating future needs as they expand rapidly. Recycling plastics presents greater challenges, as many programs struggle to achieve the required quality standards. The recycling of Poly Vinyl Chloride frequently leads to downcycling, resulting in the production of items that meet lower quality benchmarks.

1.2.3 Aluminium recycling provides significant benefits:

1. Approximately 75% of aluminium ever produced is still in use today as it can be recycled endlessly without compromising its properties or quality.
2. Recycling aluminium saves 95% of the energy it would take to make new aluminium metal.
3. The recycling process includes sorting, shredding, cleaning, melting, removal of byproducts, and creation of aluminium alloy as shown below figure 6.

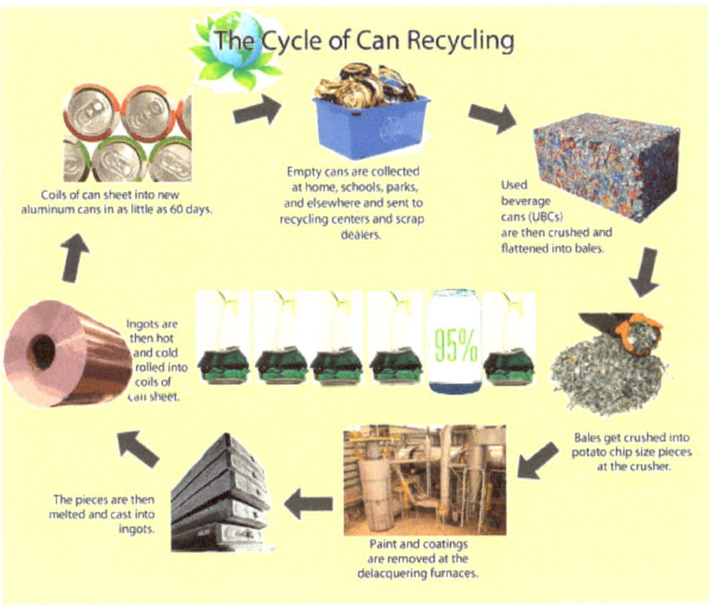

Figure 6 Aluminium recycling Process

Textile recycling involves the following steps:

Retrieving fibres, yarns, or fabrics and transforming them into new, valuable products. 1 Classifying used textiles based on their material composition and colour. Breaking down or shredding these textiles into raw fibres. Carefully cleaning the fibres. Re-spinning the cleaned fibres into new textile products. Repurposing them to create rags, clothing, insulation, and various other items as shown in figure 7.

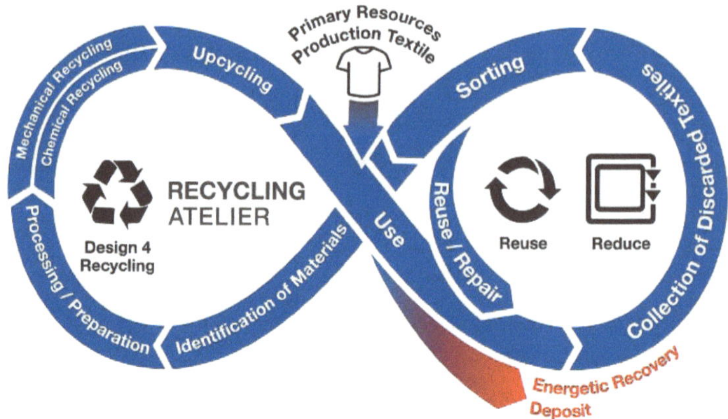

Figure 7 Textile waste recycling process

1.2.4 Paper recycling

The process of recycling paper involves converting discarded paper into new paper products. This practice offers several significant advantages: it prevents waste paper from cluttering homes and generating methane as it decomposes. Since paper Fibers contain carbon—originally absorbed from the trees used to create them—recycling helps to sequester this carbon for an extended period, keeping it out of the atmosphere. Currently, approximately two-thirds of all paper products in the United States are recovered and recycled, although not all of it is transformed into new paper. With repeated processing, the Fibers become too short for the production of new paper, which is why virgin fibers from sustainably managed forests are often incorporated into the pulp mixture as shown in figure 8.

Waste sorting

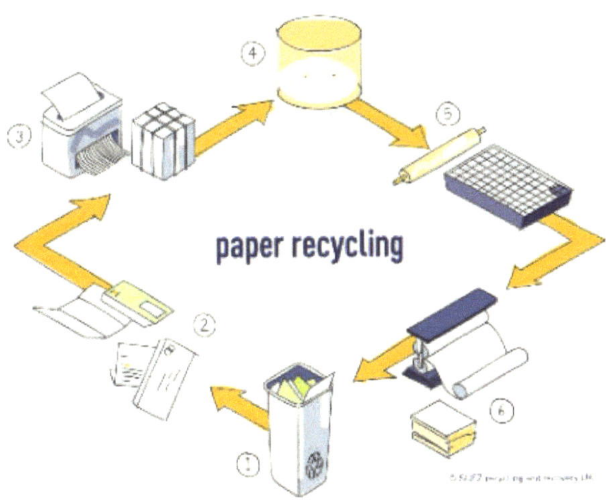

Figure 8 Paper recycling process

1.2.5 Waste sorting

Waste sorting refers to the practice of dividing waste into various components as shown in figure 9. This process can be carried out manually at the household level or through curb side collection programs, as well as automatically in materials recovery facilities or mechanical biological treatment systems. Historically, hand sorting was the initial technique employed in waste sorting. Additionally, waste can be sorted at civic amenity sites. Waste segregation involves categorizing waste into dry and wet types. Dry waste encompasses materials such as wood, metals, and glass, while wet waste generally pertains to organic refuse, often produced by food establishments, and tends to be heavier due to moisture content. Segregation occurs at the point of disposal or collection, ensuring that each type of waste is placed in its respective category, whereas sorting takes place after disposal or collection. Effective waste segregation leads to the production of high-quality, pure materials, while sorting may result in lower-quality, contaminated materials. Currently, automatic waste segregation systems are becoming increasingly popular and are already in use in various regions around the globe, including Australia.

Figure 9 Waste sorting

1.2.6 Electronic waste

Commonly known as e-waste, encompasses discarded electronic devices that are no longer functional, including computers, smartphones, and televisions. This category of waste is the fastest-growing globally and poses significant environmental challenges if not managed appropriately. To tackle the issue of e-waste, several recycling initiatives have been established, including: - National recycling programs in Australia that accept a diverse array of electronic devices. - In-store recycling services provided by retailers for small technology items and batteries. - Community-driven efforts aimed at encouraging the reuse and recycling of e-waste, which aid in conserving essential resources and minimizing landfill contributions. For further details on recycling e-waste in your locality, you may refer to local resources such as the Brisbane City Council as shown in figure 10.

Figure 10 Electronic waste

1.2.7 Plastic recycling

Plastic recycling involves converting plastic waste into new products. This process can help decrease reliance on landfills, conserve natural resources, and mitigate environmental harm caused by plastic pollution and greenhouse gas emissions. However, recycling rates for plastics are significantly lower compared to other recyclable materials like aluminum, glass, and paper. Since the inception of plastic production until 2015, approximately 6.3 billion tonnes of plastic waste have been generated globally, with only 9% being recycled and around 1% undergoing multiple recycling processes. Among the remaining waste, 12% was incinerated, while 79% ended up in landfills or contributed to environmental pollution. The majority of plastics are non-biodegradable, and in the absence of recycling efforts, they disperse throughout the environment, leading to plastic pollution. For instance, by 2015, an estimated 8 million tonnes of plastic waste entered the oceans each year, adversely affecting marine ecosystems and creating large areas of ocean debris as shown in figure11.

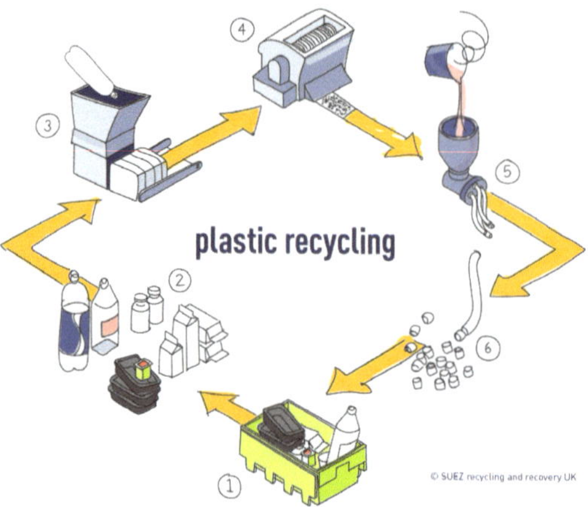

Figure 11 Plastic recycling

1.3 Questions and explanations

1. **The term 'Reduce' refers to**

   The term 'Reduce' refers to using less of something, creating smaller amounts of waste. This means minimizing the consumption of resources and minimizing the generation of waste in order to protect the environment and conserve resources.

2. **Reducing, Reusing, and Recycling make the Earth cleaner.**

   Reducing, reusing, and recycling are sustainable practices that help in minimizing waste and conserving resources. By reducing the amount of waste we produce, reusing items instead of throwing them away, and recycling materials, we can reduce pollution, save energy, and protect natural habitats. These actions contribute to making the Earth cleaner and more sustainable for future generations.

3. **The term 'reuse' refers to:**

   The term 'reuse' refers to using something over and over again. This means that instead of disposing of an item after one use, it is used multiple times, reducing the need for producing or purchasing new items. Reusing helps to conserve resources, reduce waste, and minimize the environmental impact of consumption.

4. **Plastic bags are better for the environment than reusable bags. State true or false.**

   Plastic bags are not better for the environment than reusable bags. Reusable bags are more environmentally friendly because they can be used multiple times, reducing the amount of waste generated. Plastic bags, on the other hand, are often used once and then discarded, contributing to pollution and litter. Additionally, plastic bags are made from non-renewable resources like petroleum, while reusable bags can be made from sustainable materials like cotton or recycled materials. Therefore, the statement that plastic bags are better for the environment is false.

5. **The term 'recycle' means:**

   The term 'recycle' refers to the process of transforming waste materials into new products or materials. It involves taking something that is no longer useful or needed and converting it into a new item that can be used again. This process helps to reduce waste and conserve resources by giving a new life to materials that would otherwise end up in landfills.

6. **Recycling is something only adults can do.**

   The statement "Recycling is something only adults can do" is incorrect. Recycling is an activity that can be done by people of all ages, including children. It is important to teach children about the importance of recycling and involve them in the process to instill good environmental habits from a young age.

7. **Which of the following is NOT a recyclable material?**

   Food is not a recyclable material because it decomposes and cannot be processed through traditional recycling methods. Paper, plastic, and metal can all be recycled through various processes to create new products.

8. **You can reduce your water usage by shutting off the water while you brush your teeth and taking shorter showers. State true or false.**

   By shutting off the water while brushing your teeth and taking shorter showers, you can reduce your water usage. This is because when the water is running while brushing teeth, a significant amount of water is wasted unnecessarily. Similarly, taking shorter showers reduces the amount of water used compared to longer showers. Therefore, the statement is true.

9. **Which of the following is bad for the environment?**

   Littering is bad for the environment because it involves the careless disposal of waste materials in public spaces, such as throwing trash on the ground or in water bodies. This leads to pollution, as litter can contaminate soil, water, and air, harming plants, animals, and ecosystems. Littering also contributes to the spread of diseases and can be hazardous to humans and wildlife. It is important to promote responsible waste management practices like recycling, reducing waste, and reusing items to minimize the negative impact on the environment.

10. **No garbage can be recycled. True or false?**

    False. This statement is incorrect because garbage can indeed be recycled. Recycling involves the process of converting waste materials into reusable materials, reducing the amount of waste sent to landfills. Various items such as paper, plastic, glass, and metal can be recycled to conserve resources and protect the environment.

11. **How much energy could 1 recycled tin can save?**

    Recycling a tin can can save a significant amount of energy. However, the amount of energy saved may not be enough to power a TV for 6 hours or a cell phone for 18 hours. The correct answer states that recycling 1 tin can can save enough energy to power a TV for 3 hours, which suggests that recycling tin cans can have a significant impact on energy conservation.

12. It takes more energy to create paper than it does to recycle paper. True or false?

It is true that it takes more energy to create new paper from raw materials than it does to recycle paper. The process of creating paper from scratch involves cutting down trees, transporting them, and using various chemicals and energy-intensive machinery to break down the wood fibers and create new paper. On the other hand, recycling paper involves collecting used paper, removing any contaminants, and then reprocessing it into new paper. This recycling process requires less energy and resources compared to the production of new paper, making it a more energy-efficient option.

2.1. The Five R's: The Basics of Sustainability

Let us revisit a topic that you are probably familiar with: reduce, reuse, and recycle. Now let us add rethink and refuse to that list, and there you have the basics of sustainability as Shon in figure 12. The five R's: reduce, reuse, recycle, rethink, and refuse are a great way to implement small changes into your daily life and live more environmentally friendly and conscious.

Figure 12 Five R's: The Basics of Sustainability

Rethink:

The term "rethink" encompasses the broader concept of reassessing your consumption patterns to reduce waste. Here are some considerations to help you adopt a lifestyle that fosters a sustainable future. Is the purchase of this product justified by the waste or environmental damage it may cause? Do I truly love this item or genuinely need it? Will I still be using it in six months? Am I willing to take the necessary steps to dispose of these items responsibly rather than simply discarding them? What happens to my waste, and where does it ultimately end up? Could this item serve another purpose before I throw it away?

Refuse:

The act of "refusing" involves declining items that do not align with your commitment to a sustainable future. Reject fast fashion brands. If you cannot find what you need second-hand, consider seeking out eco-friendly brands. By supporting these brands, you are making a choice for sustainability with your spending. Politely decline offers of single-use items when you are out, unless they are absolutely necessary. Avoid supporting practices that do not promote a more sustainable lifestyle. There are many additional actions you can take to live in a more environmentally conscious manner, but these suggestions can serve as a starting point as you strive to be more eco-friendly!

3.1. The Current National Picture

The Environmental Protection Agency (EPA) has been gathering and disseminating data regarding waste generation and management in the United States for over 35 years. This data is utilized by the Agency to evaluate the effectiveness of materials management initiatives nationwide and to analyze the national waste stream. The statistics presented are up to date as of the end of the 2018 calendar year. In 2018, the total amount of municipal solid waste (MSW) generated reached 292.4 million tons (measured in U.S. short tons, unless otherwise noted), which translates to an average of 4.9 pounds per person per day.

Out of this total, around 69 million tons were recycled, and 25 million tons were composted. In total, nearly 94 million tons of MSW were either recycled or composted, resulting in a recycling and composting rate of 32.1 percent. Additionally, 17.7 million tons of food waste were managed through various alternative methods, including animal feed, bio-based materials/biochemical processing, co-digestion/anaerobic digestion, donations, land application, and sewer/wastewater treatment. For further details on food waste management, please refer to Food: Material-Specific Data shown in figure 13. Furthermore, close to 35 million tons of MSW (11.8 percent) were incinerated with energy recovery, while over 146 million tons of MSW (50 percent) were disposed of in landfills.

The Environmental Protection Agency (EPA) defines municipal solid waste (MSW) as the assortment of items that consumers discard after use. This category encompasses a variety of materials, including bottles, cardboard boxes, food waste, grass clippings, furniture, computers, tires, and refrigerators. However, it is important to note that MSW does not encompass all materials that may be disposed of in landfills at the local level, such as construction and demolition (C&D) debris, municipal wastewater sludge, and other types of non-hazardous industrial waste. Although the analysis presented in Facts and Figures primarily concentrates on MSW, the EPA has recently begun to include estimates of C&D generation and management as a distinct non-hazardous waste category.

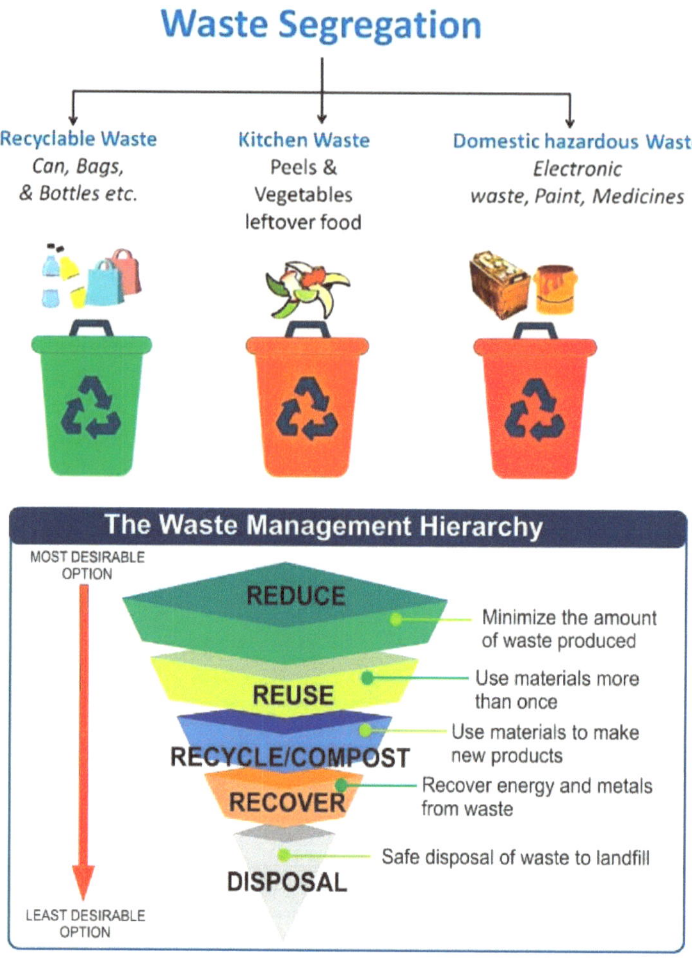

Figure 13 waste management process

EPA is now in the process of reviewing the waste hierarchy to determine if potential changes should be made based on the latest available data and information.

4.1 Hazardous Materials and Waste Management Hierarchy

EPA developed the non-hazardous materials and waste management hierarchy in recognition that no single waste management approach is suitable for managing all materials and waste streams in all circumstances.

The hierarchy ranks the various management strategies from most to least environmentally preferred. The hierarchy places emphasis on reducing, reusing, recycling and composting as key to sustainable materials management. These strategies reduce greenhouse gas emissions that contribute to climate change.

Source Reduction and Reuse

Source reduction, often referred to as waste prevention, involves minimizing waste generation at its origin and is regarded as the most environmentally favourable approach. This strategy can manifest in various ways, such as reusing or donating items, purchasing in bulk, minimizing packaging, redesigning products, and decreasing toxicity levels. Additionally, source reduction plays a crucial role in manufacturing, with trends like lightweight packaging, reuse, and remanufacturing gaining traction in the business sector. Opting for products that feature these characteristics promotes source reduction. The benefits of source reduction include:

- Decreasing greenhouse gas emissions that contribute to climate change,
- Preserving natural resources,
- Conserving energy,
- Mitigating pollution,
- Lowering the toxicity of waste, and
- Providing cost savings for both consumers and businesses.

Recycling and Composting

Recycling encompasses a range of processes that involve the collection of items that are used, reused, or deemed waste, followed by the sorting and processing of these recyclable materials into raw resources, and ultimately remanufacturing these recycled materials into new products. The final step in the recycling chain is completed by consumers who buy products made from recycled materials.

Additionally, recycling can involve composting organic waste such as food scraps and yard trimmings. The advantages of recycling include:

- Decreasing greenhouse gas emissions that contribute to climate change;
- Preventing the release of air and water pollutants;
- Conserving energy;
- Providing essential raw materials for industry;
- Generating employment opportunities;
- Encouraging the advancement of environmentally friendly technologies;
- Preserving resources for future generations; and
- Minimizing the necessity for new landfills and incinerators.

Energy Recovery

The process of energy recovery from waste involves transforming non-recyclable waste materials into usable forms of energy such as heat, electricity, or fuel. This transformation is achieved through various methods, including combustion, gasification, pyrolysis, anaerobic digestion, and the recovery of landfill gas (LFG). Commonly referred to as waste-to-energy (WTE), this approach not only produces a renewable energy source but also helps lower carbon emissions by reducing reliance on fossil fuels and minimizing methane production from landfills. Following the energy recovery process, about ten percent of the original waste volume remains as ash, which is typically disposed of in a landfill.

Treatment and Disposal

Prior to disposal, treatment can help reduce the volume and toxicity of waste. Treatments can be physical (e.g., shredding), chemical (e.g., incineration), and biological (e.g., anaerobic digestor). Landfills are the most common form of waste disposal and are an important component of an integrated waste management system. Modern landfills are well-engineered facilities located, designed, operated, and monitored to ensure compliance with state and federal regulations. Landfills that accept municipal solid waste are primarily regulated by state, Tribal, and local governments. EPA, however, established national standards that these landfills must meet in order to stay open. The federal landfill regulations eliminated the open dumps (disposal facilities that do not meet federal and state criteria) of the past. Today's landfills must meet stringent design, operation, and closure requirements. Methane gas, a byproduct of decomposing waste, can be collected and used as fuel to generate electricity. After a landfill is capped, the land may be used for recreation sites such as parks, golf courses, and ski slopes.

5.1 Definition, Scope and Importance of Environment

1. **Definitions**

   Environment: The environment encompasses the entire physical and biological system in which humans and other organisms exist. Environmental studies address all issues that impact living beings.

   Environmental Science: Environmental science is the discipline that examines the workings of nature and the interrelationships among various natural elements. The environment comprises several interacting components, including biology, geology, chemistry, physics, engineering, sociology, health, and economics. Effective and realistic planning is essential to achieve a balance among these elements. Consequently, environmental science adopts a multidisciplinary approach.

2. **Scope:** Our natural landscape includes forests, rivers, deserts, rocks, minerals, and soil. Human activities transform these landscapes into villages, towns, or cities. Despite residing in urban areas, we rely on agricultural produce from surrounding villages, highlighting the connection between our daily lives and the environment. We require water

and air for survival, as well as other resources integral to our daily activities. Humans are heavily reliant on nature and the environment. Therefore, it is crucial to conserve environmental resources such as water, trees, minerals, food, energy, and land in their natural state. In contemporary times, technological advancements have enabled increased food production using fertilizers and pesticides, while the construction of dams has contributed to environmental degradation. Many environmental resources, including water, minerals, petroleum products, and timber, are being continuously extracted. Ecologists and environmental scientists have recognized that this pattern of consumption will lead to the degradation and depletion of the natural environment. A key feature of numerous environmental resources is their non-renewable nature: once a natural resource is depleted, it cannot be reproduced in its original form. Continued extraction may result in their unavailability for future generations.

3. **Significance:**

Environmental studies adopt a multidisciplinary approach. Environmental resources serve multiple functions and are subject to market pricing. Scarce natural resources, which are limited in availability, tend to be more expensive; for instance, wood and water. • A significant volume of clean water from nature is being contaminated and squandered. Byproducts from chemical processes contribute to water pollution, while gases released into the atmosphere degrade air quality. • Deforestation, or the removal of trees, results in rising environmental temperatures, drying rivers, and a lack of fresh air. The cumulative impact of these factors creates an unhealthy atmosphere for humans, leading to various health issues. • Raising awareness about the importance of preserving nature and the environment can help mitigate the misuse or waste of natural resources. It is essential for everyone to participate in protecting the environment and preventing ecological harm

6.1 The Importance of Public Awareness

The depletion of the Earth's natural resources is occurring at an alarming rate, and human activities are contributing significantly to environmental degradation. We cannot rely solely on governments to protect the environment. It is essential for each individual to take on a crucial role in this effort. A practical step we can take is to minimize the waste of natural resources.

a) Individuals and the Environment

Throughout history, individuals have shown concern for the environment, shaped by their perceptions of environmental challenges and their attitudes towards them. • The environmental movement has broadened its focus to encompass all facets of the natural world, including land, water, minerals, living organisms, biological processes, the atmosphere, climate, oceans, and even outer space. • This movement has also extended its analysis to include international economic cooperation, addressing issues such as commodity pricing, structural adjustments, and subsidies related to products derived from natural resources. • Numerous environmentalists have shared their insights to advocate for the protection of nature, wildlife, ecosystems, agriculture, and environmental legislation. Notable figures include Charles Darwin, Salim Ali, Indira Gandhi, S.P. Godrej, Madhav Gadgil, M.C. Mehta, Medha Patkar, and Sundarlal Bahuguna. • Individuals can engage in environmental advocacy by forming pressure groups, serving as watchdogs, participating in advisory councils, and reinforcing environmental regulations.

b) Institutions in the Environment / NGOs

Scientific organizations and Non-Governmental Organizations (NGOs) have significantly contributed to the environmental movement. • Environmental organizations encompass a diverse array of interests. Smaller groups are formed to address local issues, while others focus on specific concerns at a national level.

• In recent decades, numerous international environmental NGOs have emerged, including influential entities such as Friends of the Earth, Greenpeace, and the World Wide Fund for Nature (WWF). These groups enable individuals to impact both national and international policies. • The media serves as a powerful tool for raising public awareness about various environmental issues; however, it tends to be more reactive than proactive

c) Environmental Education

The study and education of environmental issues are crucial for the preservation of our surroundings.

Scope of Environmental Studies 1. Environmental studies foster awareness and sensitivity towards the entire ecosystem and the challenges it faces. 2. They encourage active participation in the protection and enhancement of the environment. 3. These studies cultivate the skills necessary to identify and address environmental issues. 4. They emphasize the importance of conserving natural resources. Importance of Environmental Studies 1. They facilitate an understanding of the principle of "development that does not harm the environment." 2. They provide insights into various environmental hazards. 3. They promote awareness and advocacy for environmental protection laws and their enforcement. 4. They help connect the quality of life to environmental conditions.

7.1 Structure of Atmosphere and Functions of Atmosphere

The Earth's surrounding gaseous layer is referred to as air or the atmosphere. The physical and chemical composition of the atmosphere is organized into five concentric layers.

These layers are :

1. Troposphere (10 km above earth)

2. Stratosphere (18 to 50 km)
3. Mesosphere (51 to 85 km)
4. Thermosphere / Ionosphere (upto 500 km)
5. Exosphere

1. Troposphere:

- The troposphere is the atmospheric layer closest to the Earth's surface, ranging from 10 to 18 kilometers in altitude. - This layer holds 75% of the total atmospheric mass and is also rich in moisture. - Within the troposphere, temperatures gradually decrease from 15 °C to -56 °C. - The primary chemical components found in the troposphere include $O2$, $CO2$, $H2O$, and $N2$. 2. Stratosphere:

- The stratosphere lies above the troposphere, extending from 18 to 50 kilometers above the Earth's surface. - It is characterized by a temperature range between -55 °C and 5 °C. - The stratosphere is abundant in ozone ($O3$) and is devoid of moisture and clouds. - This layer plays a crucial role in shielding the Earth from the sun's ultraviolet radiation.

3. Mesosphere:

- The mesosphere spans altitudes of 50 to 85 kilometers above the Earth's surface. - This layer is characterized by low temperatures, reaching as low as -92 °C, along with diminished atmospheric pressure. - The mesosphere primarily consists of nitrogen ($N2$) and contains minimal amounts of ozone gas.

4. Thermosphere: - The thermosphere reaches heights of approximately 450 to 500 kilometers above the Earth's surface. - In this layer, temperatures gradually rise, potentially reaching up to 1200 °C. - The thermosphere is composed of charged particles, including $O+2$, $O+$, and $NO+$. 5. Exosphere:

- The exosphere extends from 1600 kilometers to 3000 kilometers above the Earth's surface. - This layer is characterized by extremely high temperatures due to direct exposure to solar radiation. - The exosphere is primarily made up of hydrogen (H-) and helium (He).

Functions of the Atmosphere

- The atmosphere serves several crucial functions, including: 1. The absorption of infrared (IR) radiation by the atmosphere, which helps to regulate the Earth's heat balance. 2. The presence of various gases, such as oxygen, carbon dioxide, and nitrogen, within the atmosphere, which are essential for supporting life on Ear

8.1 Multiple Choice Questions on Recycling

1. Recycling reduces

   (A) energy usage

   (B) air pollution

   (C) water pollution

   (D) all of the above

2. The international recycling logo consists of ___ chasing arrows.

   (A) two

   (B) three

   (C) four

   (D) five

3. Recycle is ___ component of waste hierarchy.

   (A) first

   (B) second

(C) third

(D) fourth

4. Following ISO standard is for environmental management control of recycling practice.

   (A) 14001:2012

   (B) 14001:2013

   (C) 14001:2014

   (D) 14001:2015

5. Which of the following legislative option(s) has (have) used to create stable supply?

   (A) mandatory recycling collection

   (B) container deposit legislation

   (C) refuse bans

   (D) all of the above

6. Conatiner deposit legislation involves

   (A) imposing fine for not returning the container

   (B) offering a refund for the return of certain containers

   (C) crushing the container after use

   (D) none of the above

7. The raw material that is sent to, and processed in a waste recycling plant or materials recovery facility which will be used to form new products

   (A) Recyclate

   (B) Recycling material

   (C) recycled material

   (D) none of the above

8. Following material has highest recyclate quality

    (A) paper

    (B) plastic

    (C) steel

    (D) oil

9. Following is (are) the system(s) to collect recyclates from the general waste stream.

    (A) Curbside collection

    (B) buy back centers

    (C) drop-off centers

    (D) all of the above

10. Plastic Pyrolysis can convert petroleum based streams such as plastics into

    (A) fuels

    (B) carbons

    (C) both (A) and (B)

    (D) wax

11. Following is (are) suitable raw material(s) for pyrolysis

    (A) Mixed plastic

    (B) Mixed waste plastic from waste paper mill

    (C) Multi-layered plastic

    (D) All of the above

12. Type 1 (polyethylene terephthalate) plastic is commonly found in

    (A) soft drinks and water bottles

    (B) laundry detergent bottle

(C) pipes

(D) shopping bags

13. Type 3 (polyvinyl chloride) plastic is commonly found in

    (A) soft drinks and water bottles

    (B) laundry detergent bottle

    (C) pipes

    (D) shopping bags

14. In which of the following recycling, maximum energy is saved

    (A) steel

    (B) cardboard

    (C) paper

    (D) aluminium cans

ANSWERS:

1-(D), 2-(B), 3-(C), 4-(D), 5-(D), 6-(B), 7-(A), 8-(C), 9-(D), 10-(C), 11-(D), 12-(A), 13-(C), 14-(D)

8.2 Multiple Choice Questions on Renewable Energy Resources

1. Which of the following is (are) renewable resource(s)

   (A) wind

   (B) tides

   (C) geothermal heat

   (D) all of the above

2. Which of the following country generate all their electricity using renewable energy?
   (A) Iceland
   (B) England
   (C) USA
   (D) China
3. Renwable energy often displaces conventional fuel in which of the following area
   (A) space heating
   (B) transportation
   (C) electricity generation
   (D) all of the above
4. Which of the following is used as fuel for transportation
   (A) ethanol
   (B) aldehyde
   (C) ketone
   (D) all of the above
5. Biodiesel is produced from oils or fats using
   (A) fermentation
   (B) transesterification
   (C) distillation
   (D) none of the above
6. Photovoltaic cell converts solar energy into
   (A) heat energy
   (B) electric energy

(C) mechanical energy

(D) chemical energy

7. In which of the following region winds are stronger and constant

   (A) deserts

   (B) offshore

   (C) low altitudes sites

   (D) all of the above

8. Following country met more than 40% of its electricity demand from wind energy

   (A) Denmark

   (B) Portugal

   (C) Ireland

   (D) Spain

9. Concentrated solar power (CSP) systems use ____ to focus a large area of sunlight into a small beam.

   (A) lenses

   (B) mirrors

   (C) tracking systems

   (D) all of the above

10. The difference, in temperature between the core of the planet and its surface, is known as

    (A) geothermal coefficient

    (B) geothermal gradient

    (C) geothermal constant

    (D) none of the above

11. Biomass can be converted to

    (A) methane gas

    (B) ethanol

    (C) biodiesel

    (D) all the above

12. The International Renewable Energy Agency (IRENA) was formed in

    (A) 2008

    (B) 2009

    (C) 2010

    (D) 2011

13. Which of the following was the first solar powered aircraft to complete a circumnavigation of the world?

    (A) Solar impulse

    (B) Solar impulse 2

    (C) Solar impulse 3

    (D) Solar impulse 4

14. Following is true for biomass and biofuels

    (A) their contribution in reduction in CO_2 emissions is limited

    (B) both emit large amount of air pollution when burned

    (C) they consume large amounts of water

    (D) all of the above

ANSWERS:

1-(D), 2-(A), 3-(D), 4-(A), 5-(B), 6-(B), 7-(B), 8-(A), 9-(D), 10-(B), 11-(D), 12-(B), 13-(B), 14-(D)

Chapter-2

Ecosystems & Environmental Sustainability Solutions

1.1 Ecosystems (Structure and Function)

Ecology is study of interactions among organisms with their environment. The environment consists of both biotic components (living organisms) and abiotic components (non-living organisms). The terms ecosystems is combination of two words, where 'eco' implies the environment and 'system' implies an interacting, inter-dependent complex.

1.2 Definitions of Ecosystem

An ecosystem is a group of plants and animals along with physical environment with which it interacts. An ecosystem is a community of different species interacting with one another and with their environment exchanging energy and matter. An ecosystem is a community of living organisms (plants, animals, and microbes) in conjunction with the non-living components of their environment (things like air, water, and mineral soil), interacting as a system. An ecosystem is a biological community of interacting organisms and their physical environment. Example: Grassland ecosystem, aquatic ecosystem, dessert ecosystem etc.

1.3 Scope and Importance of Ecosystem

Scope of Ecosystem

Our natural landscape consists of forest, river, desert, rocks, minerals and soil. These landscapes are transformed into villages, town, or cities by human beings. Even though we live in cities, the food grains are supplied from villages surrounding. It means that our daily life is linked with surrounding environment. We need water, air to survive and other day-to-day activities which are part of environment. Human beings are greatly depending on nature or environment. Therefore, environmental resources like water, trees, minerals, food, energy, land must be preserved in their natural form. Now a days, because of technological advancements more foods can be grown by using fertilizers and pesticides, also construction of dams leads to environmental degradation. Most environmental resources like-water, minerals, petroleum products, wood etc. are being extracted continuously. Ecologists and environmental scientists have recognised that if these resources are consumed in this way it will degrade and deplete natural environment. A distinguishing characteristic of many environmental resources is that they are non- producible: If the natural resource is exhausted, it is not possible to reproduce them in original form. If we continue to extract them, they may not be available for future generations.

Importance of Ecosystem

- Environmental studies involve multidisciplinary approach. Environmental resources play a multifunctional role as they command market prices.
- Any scarce natural resource (rarely available) will cost more as its supply is less e.g. wood, water.
- Huge amount of nature's clean water is being polluted and wasted. Waste by products of chemical process pollute water and gases are polluting air.
- Deforestation (cutting of trees) leads to increasing environment temperature, dry rivers, unavailability of fresh air.
- The accumulated effect of all above factors causes unhealthy atmosphere to human beings by giving variety of diseases.
- The misuse or waste of natural resources can be stopped by spreading awareness to preserve the nature or environment.

- All must contribute for safeguarding of environment and by preventing environmental damage.

1.4. Classification of Ecosystem: Based on interference and noninterference by man, there exists two types of ecosystems.

I. Natural ecosystem. 2. Artificial ecosystem.

1. **Natural ecosystem** It operates under natural condition. There is no interference by man at all. It can be divided further based on habitat.
   - i. Terrestrial ecosystem - Forest ecosystem, grassland ecosystem, dessert ecosystem.
   - ii. Aquatic ecosystem - It exists under water. It can be further divided into two types.
     - **a) Fresh water ecosystem**

       - Running water (river), stream

       - Standing water (lake, pond)
     - **b) Marine ecosystem**

       - Salt water ecosystem.
2. **Artificial / Man made ecosystem**: These ecosystems are maintained artificially by man where energy is added and manipulated through planning. Examples: Cropland, gardens aquarium etc.

1.5. Characteristics of Ecosystem

1. It is structural and functional unit of ecology.
2. Its structure is related to species diversity i.e. more complex ecosystem have high species diversity and simple ecosystem have low diversity.

3. Functions of ecosystem is related to energy flow and cycling of material involved and within ecosystem.
4. Ecosystem mature as we pass from less complex to more complex structure i.e. early stage has excess potential energy and relatively high energy flow per unit biomass than later stages. It reduces at energy stage.
5. Alterations in environment represent selective pressures upon populations to which it must adjust, those which are unable to adjust must disappear i.e. survival at fittest.
6. Environment and energy fixation in any ecosystem is limited and constant be exceeded without serious undesirable effects.

2.1 Biodiversity

The diverse forms of life on Earth have evolved over millennia to meet human needs. To comprehend the life cycles of plants and animals, it is essential to classify and categorize them accurately. Biodiversity, also known as biological diversity, refers to the variety and variability of living organisms within a specific environment. It encompasses all forms of life on our planet. • Biodiversity can be characterized at three interconnected levels of biological organization: genes, species, and ecosystems. • The survival, security, equality, and dignity of communities worldwide rely on the continuous functioning of natural systems that provide essential energy and nutrients. Therefore, it is crucial for all individuals to embrace ecological responsibility as shown in figure1.

Figure 1 Biodiversity classify and categorize

2.2 Importance of Biodiversity

1. Increase ecosystem productivity; each species in an ecosystem has a specific role to play.
2. Support a larger number of plant species and, therefore, a greater variety of crops.
3. Protect freshwater resources.
4. Promote soils formation and protection.
5. Provide for nutrient storage and recycling.
6. Aid in breaking down pollutants.
7. Contribute to climate stability.
8. Speed recovery from natural disasters.
9. Provide more food resources.
10. Provide more medicinal resources and pharmaceutical drugs.
11. Offer environments for recreation and tourism.

2.3 Types of Biodiversity

There are three basic types of biodiversity -

1. Genetic diversity
2. Species diversity
3. Ecosystem diversity or community diversity.

2.4 Genetic Diversity

Genetic diversity is a measure of variety available for the same genes within individual species. Genetic diversity is based on variation between genes i.e. functional units of hereditary information. The genetic variability is essential for a healthy breeding population of a species. Example: Each human being is different from all other, thousands of rice varieties are available.

2.5 Species Diversity

Species diversity is the number of different species of living things within an area. Species are regarded as populations within which gene flow occurs under natural conditions. Members of one species, do not breed freely with members of other species. Examples: Tiger, lion, teakwood, human being etc.

2.6 Ecosystem Diversity

a) Ecosystem diversity relates to the variety of habitats, biotic communities, and ecological processes in the biosphere as well as the diversity within ecosystems. Diversity can be described at several different levels and scales:

b) Functional diversity is the relative abundance of functionally different kinds of organisms Community diversity is the number sizes and spatial distribution of communities, and is sometimes referred to as patchiness (uneven quality)

c) Landscape diversity is the diversity of scales of patchiness.

d) No simple relationship exists between the diversity of an ecosystem and ecological processes such as productivity, hydrology and soil generation.

e) Neither does diversity correlate neatly with ecosystem stability, nor its resistance to disturbance and its speed of recovery.

f) There is no simple relationship within any ecosystem between a change in its diversity and the resulting change in the system's processes.

g) For example, the loss of a species from a particular area or region (local extinction or extripation) may have little or no effect on net primary productivity of competitors take its . place in the community.

h) The converse may be true in other cases. For example, if herbivorous such as zebra and wild beasts are removed from the African savanna, net primary productivity of the ecosystem decreases.

3.1 Environmental sustainability solutions

Global warming is an urgent global challenge that demands sustainable solutions to mitigate its devastating impacts on our planet. Global warming refers to the long-term increase in the Earth's average temperature due to human activities, primarily greenhouse gases like carbon dioxide and methane in the atmosphere as shown in figure 2. This trapped heat causes shifts in climate patterns, resulting in rising sea levels, extreme weather events, and other adverse impacts on ecosystems and the environment. Global warming is a critical issue that requires urgent action to mitigate its consequences and transition toward more sustainable practices.

As human activities continue to release greenhouse gases into the atmosphere, the Earth's temperature rises, leading to severe climate changes. To tackle this crisis, we must embrace sustainable practices and implement innovative strategies that promote a low-carbon future. This article will discuss 10 sustainable solutions for global warming and pave the way for a more resilient and environmentally friendly world.

Figure 2 Global Warming

3.2 The Impact of Global Warming on the Environment and Society

Global warming is increasing day by day, leaving harsh impacts on our environment and the society we live in. Some of its impacts are:

3.2.1 Rising Sea Levels and Coastal Erosion:

Global warming has accelerated the melting of polar ice caps and glaciers, contributing to rising sea levels. As a result, low-lying coastal regions are at risk of inundation, leading to the displacement of communities and the loss of vital ecosystems.

Sea Level Rise & Coastal Erosion: Understanding the Impacts and Solutions

Coastal regions have always been dynamic, shaped by natural forces like tides, waves, and storms. However, with the accelerating sea level rise, these areas are facing unprecedented challenges. One of the most concerning effects of rising sea levels is coastal erosion, where shorelines gradually lose land to the ocean. This process isn't just a matter of losing beaches; it affects entire ecosystems, local economies, and even cultural heritage sites.

Connection Between Sea Level Rise and Coastal Erosion

To grasp the full picture, it's important to understand the relationship between sea level rise and coastal erosion. As sea levels increase, driven largely by melting ice caps and glaciers and the expansion of seawater as it warms, waves reach further inland. This not only washes away beaches but also leads to the erosion of cliffs, dunes, and other natural barriers. Over time, this can result in significant land loss, putting homes, infrastructure, and entire communities at risk.

Coastal erosion isn't just caused by rising seas; other factors also play a role. Natural processes like currents, waves, and weather patterns constantly shape coastlines. However, human activities such as construction, dredging, and deforestation can exacerbate these effects, making some areas more vulnerable than others shown in figure3.

What does the future hold for coastal regions as sea levels continue to rise? Experts predict that coastal erosion rates could increase dramatically by 2050, especially in vulnerable areas. Coastal cities in Spain, such as Barcelona, Valencia, and Malaga, are particularly at risk due to their dense populations and economic activities centered around their coastlines

Figure 3 Sea Level Rise on Coastal Areas

Case Studies of Vulnerable Coastal Regions

Take, for example, the Ebro Delta in Spain—a prime example of a coastal region already grappling with the impacts of sea level rise. This unique delta, known for its rich biodiversity and wetlands, has been experiencing alarming rates of erosion. Predictions suggest that without intervention, large parts of the delta could be underwater within decades.

Other global examples, such as the disappearing coastlines in Louisiana, USA, and the erosion hotspots in the Maldives, provide further evidence of the impending crisis. These cases show the importance of taking swift action to mitigate the impacts of sea level rise and protect coastal communities.

Adaptation and Mitigation Strategies for Coastal Communities

While the threat of rising seas and erosion is daunting, there are several adaptation and mitigation strategies that can help coastal communities better prepare for the future.

Government Initiatives and Policies to Combat Coastal Erosion

Governments worldwide, including in Spain, are increasingly recognizing the need to invest in coastal defense mechanisms. From building seawalls and breakwaters to implementing shoreline management programs, these initiatives aim to reduce the impact of waves and prevent further erosion. Spain's Ministry for Ecological Transition has already started various projects to protect vulnerable coastlines, combining both hard and soft engineering solutions.

Innovative Technology Solutions for Protecting Coastlines

Innovation also plays a critical role in addressing these challenges. Solutions like beach nourishment, which involves adding sand to eroding beaches, and living shorelines, which use natural materials like plants and rocks to stabilize the coast, are gaining popularity. Additionally, technologies such as erosion 1 methods and climate adaptation strategies help communities stay resilient against the relentless advance of the sea as shown in figure 4.

Figure 4 Role of Climate Change - Coastal Erosion

It's impossible to discuss sea level rise and coastal erosion without addressing the elephant in the room: climate change. The rise in global temperatures due to greenhouse gas emissions is directly linked to melting polar ice and thermal expansion of seawater. This not only leads to higher sea levels but also intensifies extreme weather events like hurricanes and typhoons, which can cause severe erosion in coastal areas.

Impact of Extreme Weather Events on Coastal Erosion

When powerful storms strike, they can accelerate erosion, removing vast amounts of sand and soil from the coast in a short period. For instance, recent studies highlight how storms like Gloria in Spain have caused catastrophic damage to Mediterranean beaches, amplifying the urgency for climate resilience in coastal areas.

Community Engagement and Education on Coastal Resilience

Building resilience isn't just about physical defines; it also involves educating and engaging local communities. Empowering citizens with knowledge about sea level rise and coastal erosion can lead to more sustainable practices and proactive measures. Schools, local councils, and environmental organizations can work together to create awareness programs that highlight both the causes and solutions to these pressing issues.

Efforts to Involve Local Communities in Coastal Protection Measures

In many regions, community-driven efforts are proving to be effective in combating erosion. Initiatives like dune restoration, planting of native vegetation, and community clean-up drives not only protect the coastline but also foster a sense of ownership and responsibility among locals.

Conclusion

The challenge of sea level rise and coastal erosion is a complex one that requires a multi-faceted approach. It involves scientific research, government policy, innovative engineering solutions, and strong community involvement. By understanding the link between rising seas and erosion and taking decisive actions, we can protect our coastlines and ensure that they remain vibrant, thriving spaces for future generations.

3.2.2 Extreme Weather Events:

Extreme weather phenomena, including hurricanes, heat waves, and wildfires, have all become more intense due to the changing climate. These occurrences can harm infrastructure, seriously ruin livelihoods, and result in fatalities.

Extreme weather events

The atmosphere reacts very rapidly to temperature rises over the land and sea. The extra heat turbo-charges the atmosphere, and excess heat and moisture is distributed around the globe. Unusual and extreme weather events are an early warning of such changes and are hard to ignore because of their damaging impacts. They damage ecosystems as well as our health and wellbeing, economies and infrastructure. In recent years, every region of the world has experienced impacts from record weather extremes: record temperatures and dangerous heatwaves, including marine heatwaves; sudden droughts; more intense storms and rainfall and damaging floods; and widespread wildfires in areas previously unaccustomed to them. For example, in 2023, high temperatures, low humidity and continuous winds caused extreme 'megafires' in Canada on an unprecedented scale, burning 18 million hectares of land – 10 million more than previous records (NASA, 2023; WWA, 2023)

Although the impacts of climate change are experienced across the world, some regions have experienced larger changes than others. Modelling shows that global warming will occur unevenly: the continents will heat up about twice as fast as the oceans, and polar regions will heat much faster than those near the equator. A global warming of 2 °C, for example, would mean that on average land surfaces would warm by 3 °C and the oceans by 1.5 °C, but more northern latitudes could warm by considerably more (WMO, 2023b). This is exactly what has been happening in recent decades. Figure 10 gives a snapshot of global warming for 2015–2019 (note that it uses a baseline of 1951–1980). Note in particular the extreme warming of the large land masses around the Arctic. By 2022, Europe had warmed by 2.3 °C above the 1850–1900 baseline – twice the global average rate – while parts of the Arctic warmed by three times the global average (WMO, 2023b).

3.2.3 Threats to Biodiversity

Global warming poses a grave threat to biodiversity, as many species struggle to adapt to rapidly changing climatic conditions. Rising temperatures and altered precipitation patterns can disrupt ecosystems, leading to habitat loss and increased risk of extinction. The core threat to biodiversity on the planet, and therefore a threat to human welfare, is the combination of human population growth and resource exploitation. The human population requires resources to survive and grow, and those resources are being removed unsustainably from the environment. The three greatest proximate threats to biodiversity are habitat loss, overharvesting, and introduction of exotic species. The first two of these are a direct result of human population growth and resource use. The third results from increased mobility and trade. A fourth major cause of extinction, anthropogenic climate change, has not yet had a large impact, but it is predicted to become significant during this century. Global climate change is also a consequence of human population needs for energy and the use of fossil fuels to meet those needs as shown in figure 5&6. Environmental issues, such as toxic pollution, have specific targeted effects on species, but they are not generally seen as threats at the magnitude of the others.

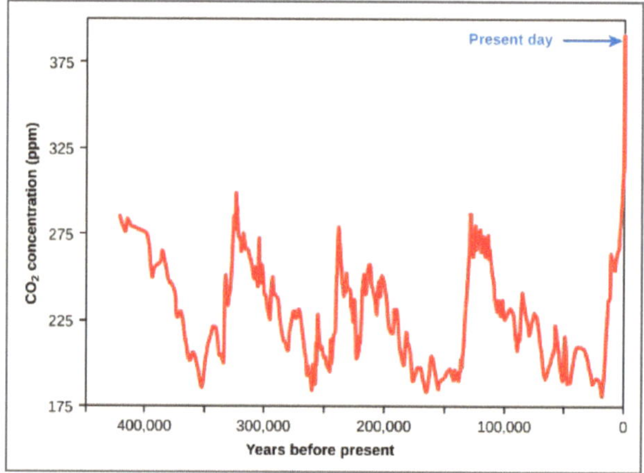

Figure 5 Threats to Biodiversity

Figure 7.3.17.3.1: Atmospheric carbon dioxide levels fluctuate in a cyclical manner. However, the burning of fossil fuels in recent history has caused a dramatic increase in the levels of carbon dioxide in the Earth's atmosphere, which have now reached levels never before seen in human history. Scientists predict that the addition of this "greenhouse gas" to the atmosphere is resulting in climate change that will significantly impact biodiversity in the coming century.

Habitat Loss

Humans rely on technology to modify their environment and replace certain functions that were once performed by the natural ecosystem. Other species cannot do this. Elimination of their ecosystem—whether it is a forest, a desert, a grassland, a freshwater estuarine, or a marine environment—will kill the individuals in the species. Remove the entire habitat within the range of a species and, unless they are one of the few species that do well in human-built environments, the species will become extinct. Human destruction of habitats accelerated in the latter half of the twentieth century. Consider the exceptional biodiversity of Sumatra: it is home to one species of orangutan, a species of critically endangered elephant, and the Sumatran tiger, but half of Sumatra's Forest is now gone. The neighbouring island of Borneo, home to the other species of orangutan, has lost a similar area of forest. Forest loss continues in protected areas of Borneo. The orangutan in Borneo is listed as endangered by the International Union for Conservation of Nature (IUCN), but it is simply the most visible of thousands of species that will not survive the disappearance of the forests of Borneo. The forests are removed for timber and to plant palm oil plantations (Figure.6). Palm oil is used in many products including food products, cosmetics, and biodiesel in Europe.

Figure 6 Threats to Biodiversity

(a) One species of orangutan, Pongo pygmaeus, is found only in the rainforests of Borneo, and the other species of orangutan (Pongo abelii) is found only in the rainforests of Sumatra. These animals are examples of the exceptional biodiversity of (c) the islands of Sumatra and Borneo. Other species include the (b) Sumatran tiger (Panthera tigris sumatrae) and the (d) Sumatran elephant (Elephas maximus sumatranus), both critically endangered species. Rainforest habitat is being removed to make way for (e) oil palm plantations such as this one in Borneo's Sabah Province. (credit a: modification of work by Thorsten Bachner; credit b: modification of work by Dick Mudde; credit c: modification of work by U.S. CIA World Factbook; credit d: modification of work by "Nonprofit Organizations"/Flickr; credit e: modification of work by Dr. Lian Pin Koh)

According to Global Forest Watch, 9.7% of tree cover was lost globally from 2002 to 2019, and 9% of that occurred in Indonesia and Malaysia (where Sumatra and Borneo are located). Figure 7. shows the average annual change in forest area around the world from 1990 to 2015. In the tropics, these losses certainly also represent the extinction of species because of high levels of endemism (species unique to those areas).

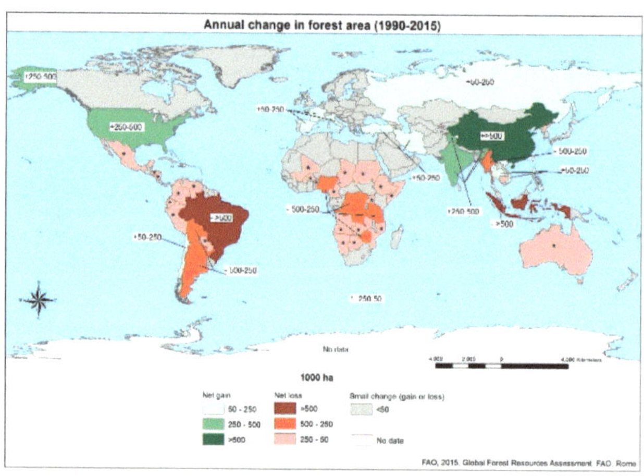

Figure 7 Average annual change in forest area

Figure 7: Average annual change in forest area globally from 2005 to 2015. China is dark green, indicating that have gained more than 500 "kilohectares" (kha) of forest. Medium green indicates countries that have gained 250-500 kha, including the the United States, India, Ghutan, and Bangladesh. Light green indicates countries that have gained 50-250 khA, including Chile, Spain,

France, Italy, Turkey, Iran, Russia, Vietnam, and Thailand. Indonesia and Brazil lost more than 500 kha of forest (marked with dark red). Countries that lost 500-250 kha are marked with medium red, including Bolivia, Argentina, Nigeria, Democratic Republic of the Congo, Tanzania, Zimbabwe, and Myanmar. Countries that lost 250-50 kha are marked with light red and an asterisk (*). These include Mexico, Honduras, Nicaragua, Venezuela, Colombia, Ecuador, Peru, Paraguay, Mali, Burkina Faso, Benin, Cameroon, Chad, Sudan, Ethiopia, Somalia, Uganda, Angola, Namibia, Botswana, Zimbabwe, Mozambique, Cambodia, North Korea, and Australia. No data has been collected for Antarctica. All other regions have had lost or gained less than 50 kha in forest area.

3.2.4 Food and Water Security Challenges:

Climate pattern shifts can disrupt agricultural productivity and affect water availability, leading to food and water shortages.

Health Impacts: Global warming contributes to the spread of infectious diseases, heat-related illnesses, and respiratory problems due to worsened air quality.

1. **The Challenges of Food Security**

   Food insecurity corresponds to a deficit in households' access to appropriate food, either in quantity and/or quality, due to limited financial resources or other factors. A related concept of nutrition insecurity is also gaining relevance, and this refers to a lack of access to food that has the adequate nutritional value necessary for a good health status and well-being [1].

   Currently, the world faces troubling challenges that greatly impact food/nutritional security. One of the most worrying trends is an increase in the world's population and the rising number of countries that are food importers owing to their inability to produce the food they need to feed their population. The United Nations World Population Prospects [2] estimates a rise in the world's population from about 8 billion in 2025 to 10.5 billion by 2100. An increase in the world's population brings the added challenge of feeding such a large number of people with the Earth's limited resources. This is only possible with more efficient food production, more sustainable use of natural resources, and the minimization and reutilization of food and agricultural waste [3].

Complying with efficient food production while minimizing environmental impacts is a major issue since the more environmentally friendly agricultural practices, such as organic farming and integrated protection, do not allow the production of food at the same rates as conventional intensive farming practices [4,5]. Nevertheless, organic farming has a decisive role in achieving sustainability in agriculture, meeting a number of the Sustainable Development Goals of the United Nations [6]. In what concerns the minimization of agricultural and food waste and its reutilization, these goals have been at the top of priorities of academics, farmers, industrials, and even consumers in attempts to minimize waste along the whole food supply chain and to find alternative ways to value foods that are discarded [7,8,9,10].

One other challenge is related to the world's food supply being strongly under pressure due to the combined effects of environmental degradation and climate change. Environmental degradation, such as sterile agricultural land, depleted water tables, and falling grain yields, threatens food security for hundreds of millions of people, especially in more vulnerable regions of the globe [3,11,12]. Climate change has direct impacts on agriculture. Climate-induced drought reduces crop production and lowers yields, while warmer temperatures lead to a higher incidence of plant, livestock, and fish diseases [13,14]. Climate change is dislocating production in some regions, and this trend is expected to increase as the century progresses. The deforestation of tropical rainforests to obtain arable soil is a pivotal issue that contributes to environmental degradation while trying to respond to a need to increase the production of higher amounts of food [15,16,17].

One other challenge is related to a trend of rising populism allied to nationalism, which has increased trade protectionism. This brings huge constraints to the commercialization of foods on a global scale [18,19]. Allied with this challenge, there is the trend for an increase in the number of weak and at-risk states which are poorly equipped to address their food insecurity and, therefore, face social and/or political instability.

2. Wars and Conflicts as Enhancers of Food Insecurity Risks

War zones and political or social conflicts are major drivers for food insecurity, not only regionally but also on a global scale. Taking the example of the Russia–Ukraine conflict, it was evident that it would have a major impact on agricultural production and trade [20]. While Ukrainian wheat, soybean, and maize production fell abruptly in 2022–2023, the global supply chain and food trade were significantly affected, causing scarcity and a consequent increase in the world's food prices of these commodities [20]. The wheat crisis could thus be alleviated by increasing production and eliminating trade restrictions [21].

At more localized scales, the consequences of wars and conflicts are devastating for the affected populations, like, for example, those derived from the Israel–Palestine conflict in 2024 [22]. Although the food security of Palestinian households in the Gaza Strip was not directly impacted by the 2014 Gaza conflict due to some resilience and adaptive capacity [23], it is also a fact that these diminish with time and with increasing intensity of the conflict, as has happened in 2024 [22].

3. Prospects and Future Challenges

Scientists and politicians have to come together to examine global food security challenges and find the correct strategies to mitigate the risks. Resilient systems must be built to better face the adverse consequences of today's still inappropriate food systems. It is a fact that much has been done so far to achieve better efficacy in food production, utilization of natural resources and minimization of food and agricultural waste, but much more needs still to be achieved for a balanced utilization of resources while producing food able to satiate the increasing world population.

Success can only be achieved through research, implementation, and collaboration in a global strategy to fight a common problem. Individualism, nationalism, and politics tend to encourage too much self-focus on one's own success and are enemies of global perspectives and collaborative approaches.

Food production, food trade, climate change, environmental degradation, food supply chains and distribution networks, waste management, governance, and food policies are all issues that will have a decisive role in the future of food security/insecurity.

3.2.5 Melting Glaciers and Permafrost

The melting of glaciers and permafrost due to global warming can release large amounts of stored carbon and methane, further exacerbating climate change.

How Melting Ice Affects People and the Planet

The mountains: From landslides and floods to droughts

From the Himalayas to the Andes to East Africa, many alpine glaciers are melting at unprecedented rates. This effect is particularly pronounced in the high mountain ranges of the tropics. Unlike peaks at the midlatitudes—say, those in the Sierra Nevadas, where heavy winter snowfalls can cause glaciers to gain mass after a summer melting period—tropical glaciers don't experience the same extreme seasonal temperature changes. As a result, they're shrinking rapidly as shown in figure 8.

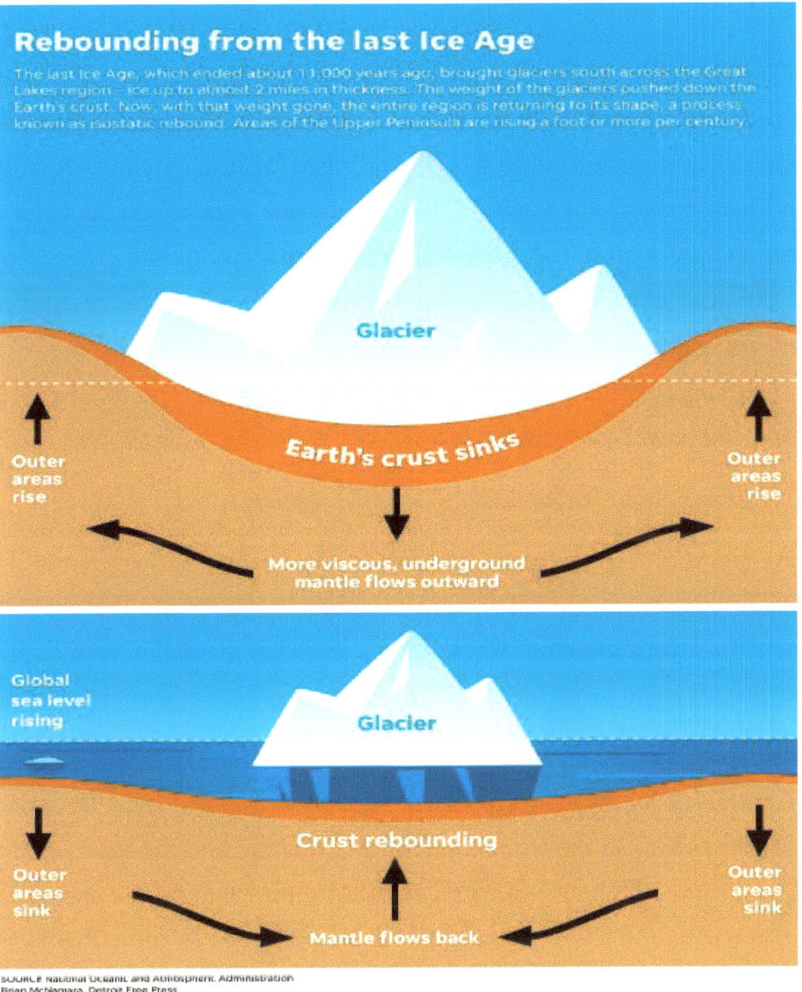

Figure 8 Melting Ice Affects People

The effects trickle down to communities in many ways. In India, areas exposed to receding glaciers resulted in a devastating landslide and flood in 2021 that killed approximately 200 people. The debris took out two hydroelectric dams and led authorities to evacuate numerousvillages downstream.

"It all started sometime around 10 in the morning. We heard a bang, which shook our village," Dinesh Negi, a resident of Raini village, told the Associated Press at the time. "We knew something wrong had happened….We could see the fury of the river." Researchers later determined that an ice and rock avalanche caused the disaster, demonstrating the increasing risk from warming and development.

In the Andes, where scientists have documented unprecedented glacier retreat, the impacts are being felt in Cordillera Blanca, a mountain range in Peru. There, local farmers who traditionally grew corn, wheat, and potatoes are contending with hotter, drier conditions. To adapt, they're experimenting with different crops, such as sugar snap peas, and relying on the increased meltwater collecting in glacial lakes. But this shift has created tensions with others seeking to profit from the lake surpluses, including large-scale irrigation projects and power companies.

The poles: From sea level rise to shellfish decline

As global average temperatures tick up, the sea ice that covers the Arctic Ocean forms later in the fall and melts earlier in the spring. That ice reflects sunlight back into the atmosphere. When it melts, the darker water surfaces below absorb more sunlight and heat—leading to further warming and more melting shown in figure 9.

Coastal sea ice also absorbs ocean wave energy. When it melts, or doesn't solidify for as long in the winter, waves crash directly into the shore, erode the land, and carve off chunks of earth. In Alaska, for example, the shorelines of many northern and western communities are shrinking by more than 3 feet a year. Several villages along the state's rivers and coasts have engaged in costly relocation efforts as they've watched their waters deluge local infrastructure.

Figure 9 Ocean wave energy

Yupik children playing near an eroding coastline in Quinhagak (Kuinerraq) village, AlaskaCredit:Mark Ralston/AFP via Getty Images

Less sea ice paired with warmer ocean temperatures could also contribute to milder cold snaps, a concern because of the rain-on-snow events that can ensue. The resulting crusty ground conditions can have dire consequences for Arctic wildlife, including reindeer herds—as well as for the communities that rely on them.A recently published study in Geophysical Research Letters suggests that "Arctic warming and sea ice loss are already impacting the Arctic communities, whose lifestyles and livelihoods were adapted to cold weather through generations of lived knowledge." That's the case for the Inuit communities of Alaska's North Slope, who can no longer rely on year-round underground ice cellars to store the meat from their hunts and ensure a consistent food supply.

Arctic sea ice loss also disrupts marine ecosystems. Researchers who investigated very cold water in the Bering Sea found that less sea ice could constrict shellfish production. The snow crab is a prime example—the fishery, estimated to be worth $227 million annually, crashed in 2018–2019. It still hasn't fully recovered, nor have the fishing communities who depend on it; a recent study by NOAA Fisheries scientists suggests that the crabs may never come back.

In Antarctica, on the other hand, the effects are quite different. Ice is now melting from below, thanks to warming seas. The impacts of climate change there are "insidiously indirect," says MacAyeal. Ice is melting "around the fringe of the continental ice sheet where it's running into the ocean." That also destabilizes ice shelves and contributes to rising seas. Scientists have been sounding the alarm bells on West Antarctica's 80-mile-wide Thwaites Glacier, which is increasingly losing structural integrity, and may very well shatter in as little as five years. The resulting chain of effects would trigger global sea level rise that impacts millions of people.

The Great Lakes: From sedimentation to algal blooms

Ice coverage across the Great Lakes, the largest freshwater system in the world, plays an important role in regulating regional weather patterns across a vast swath of the United States and Canada. But as global warming accelerates, scientists predict there will routinely be less ice atop the lakes in the winter, figure 10. The changing conditions have significant implications for local ecosystems and economies.

As the ice serves as a cap on the waters to contain water vapor from escaping into the air, less ice means more evaporation, lowering water levels. In turn, lake-effect precipitation increases, which then increases water levels. Without enough ice, the coastlines are more vulnerable to flooding and the effects of strong storms—which can include sedimentation and erosion. Researchers at Michigan State University (MSU) have found that diminishing ice cover near the shore allows waves to scour the lake bottom and transport sand from the shoreline into the lake.

Figure 10 Sedimentation to algal blooms

3.2.6 Ocean Acidification and Coral Reef Decline:

The increased absorption of carbon dioxide by the oceans leads to ocean acidification, threatening marine life, especially coral reefs.

3.2.7 Migration and Human Displacement:

Global warming-induced environmental changes can result in population displacement and migration.

India's Dynamic Migration Landscape

The Economic Survey of India 2017 estimates that the magnitude of inter-state migration in India was close to 9 million annually between 2011 and 2016The 2001 census estimated the total number of internal migrants at 314 million based on place of last residence, representing nearly 30% of the total population. According to the 2011 Census, the number of internal migrants rose to 453.6 million.According to the International Migrant Stock 2019 report (released by the Population Division of the United Nations Department of Economic and Social Affairs), India with 5 million international migrants has emerged as the top source of international migrants, constituting 6.4% of world's total migrant population.

Key Terms in Migration Dynamics

Human Migration:	It is the movement by people from one place to another with the intention of settling, permanently or temporarily in a new location (within or outside the home country). Such people are called migrants.
Immigration:	Immigration is coming to a foreign country with the intention of permanently living there.
Emigration:	Emigration is leaving a resident country with the intent to settle elsewhere.
Refugees:	These are the people who have been forced to flee their resident country because of war, violence or persecution. Such people are protected by international law, specifically the 1951 Refugee Convention.
Enroute:	Migrants in between origin and destination are known as en route.
Return Migration:	When groups of people move back to where they came from.
Seasonal Migration:	When people move with each season (e.g. farm workers following crop harvests or working in cities off-season).
World Migration Report:	It is International Organization for Migration's (IOM) flagship publication that features the latest trends in international migration, discusses emerging policy issues and provides regional recent developments in Africa, America, Asia, Europe, the Middle East and Oceania.
Migration stream and counter-stream:	A number of migrants sharing a common origin and destination form a migration stream. For every stream there is a reverse counter-stream.
International Organization for Migration (IOM):	IOM is a leading inter-governmental organization in the field of migration and works closely with governmental, intergovernmental and non-governmental partners. It was established in 1951 and has its head office at Le Grand-Saconnex, Switzerland.
Diaspora	Diaspora is commonly understood to include Non-Resident Indians (NRIs), Persons of Indian Origin (PIOs) and Overseas Citizens of India (OCI), of which PIO and OCI card holders were merged under one category – OCI – in 2015.

3.2.8 Economic Impacts

The far-reaching effects of global warming can have significant financial consequences, including damage to infrastructure, increased healthcare costs, and reduced agricultural productivity.

3.2.9 Climate Justice and Equity

Global warming disproportionately affects vulnerable populations, including low-income communities and marginalized groups.Sustainable solutions offer a way to mitigate the adverse impacts of global warming on both the environment and society. By adopting eco-friendly practices, such as renewable energy, afforestation, and sustainable agriculture, we can reduce greenhouse gas emissions, protect ecosystems, ensure resource availability, and promote climate resilience.

4.1 Sustainable Solutions for Global Warming

Sustainable Solutions for Global Warming are crucial in the contemporary world, as climate change poses unprecedented challenges. This phrase underscores the urgent need to adopt environmentally conscious practices that counteract the detrimental effects of global warming.

1. **Renewable Energy Transition**

   We must shift from fossil fuels to sustainable energy sources to reduce greenhouse gas emissions. Resources for green power, such as wind, sun, hydropower, and geothermal biofuel technologies, provide more environmentally friendly options for generating electricity. By putting money into and supporting these renewable energy sources, we can drastically reduce our carbon footprint and have less of an influence on the environment.

2. **Energy Efficiency and Conservation**

   Implementing sustainable solutions for global warming is essential to curbing its impacts. Energy efficiency and conservation play a vital role in reducing global warming. Energy-efficient technologies and practices can be implemented in buildings, homes, transportation, and industries to reduce energy use and emissions. Encouraging responsible energy use at all levels will contribute to a sustainable and green future.

3. **Afforestation and Reforestation**

   Afforestation and reforestation are critical sustainable solutions for combating global warming. Forests act as natural carbon sinks, absorbing and storing carbon dioxide. Afforestation, the establishment of new forests, and reforestation, the restoration of degraded forests, can significantly enhance carbon sequestration efforts. Government incentives and community involvement in tree planting initiatives are essential to maximizing the positive impact on the environment.

4. **Sustainable Agriculture and Land Use**

   Sustainable agriculture is an essential solution for reducing greenhouse gas emissions. The agricultural sector contributes significantly to global warming through deforestation, methane emissions from livestock, and synthetic fertilizers. Sustainable farming practices, such as agroforestry, organic farming, and efficient land use, can reduce emissions and promote soil health. Effective land management is crucial to protecting natural habitats and preserving biodiversity.

5. **Circular Economy and Waste Reduction**

   Circular economy practices play a crucial role in sustainable solutions for global warming. A circular economy can drastically cut production and disposal-related greenhouse gas emissions by reusing resources and minimizing waste. Emphasizing waste reduction, recycling, and responsible consumption will contribute to a more sustainable and eco-friendly society.

6. **Climate-Friendly Transportation**

Climate-friendly transportation is a sustainable solution to reduce greenhouse gas emissions. The transportation sector contributes significantly to global warming due to its heavy reliance on fossil fuel-powered vehicles. Investing in public transportation infrastructure, promoting active transportation options like cycling and walking, and encouraging electric transport are all crucial steps in lowering transportation-related carbon emissions.

7. **Green Building and Urban Planning**

Green building and urban planning are vital sustainable solutions for global warming. Green building practices and sustainable urban planning can lead to energy-efficient buildings and cities. Emphasizing green construction materials, passive design, and renewable energy integration can significantly reduce energy consumption. Well-planned cities with improved public transportation and green spaces foster a more sustainable urban environment.

8. **Climate Change Education and Awareness**

Climate change education is essential to promote sustainable solutions for global warming. Spreading knowledge about global warming and its effects is crucial for fostering behavioural change and encouraging sustainable practices. Climate change education programs, public campaigns, and media outreach are vital in motivating individuals and communities to take meaningful actions against climate change.

9. **Carbon Pricing and Market Mechanisms**

Carbon pricing is a sustainable solution to encourage emission reduction efforts. Implementing carbon pricing through taxes or cap-and-trade systems can incentivize industries to reduce emissions and invest in cleaner technologies. Market mechanisms like carbon offsetting and credit trading can further drive emission reduction efforts and support sustainable projects.

10. **International Cooperation and Policy Commitments**

    International cooperation is essential for effective, sustainable solutions to global warming. International collaboration and political commitments are necessary to address the collective issue of global warming. Nations must collaborate to set ambitious targets, share sustainable technologies, and provide financial support to developing countries in their climate action efforts. International agreements, like the Paris Agreement, serve as crucial frameworks to drive global climate action.

Benefits of Sustainable Solutions for Global Warming

Sustainable solutions for global warming offer various benefits that extend beyond environmental preservation. These solutions, driven by a commitment to reducing greenhouse gas emissions and curbing climate change, bring about positive transformations across various aspects of society.

5.1 Questions on General sustainability

What is sustainability?

The word sustainability comes from 'to sustain', which essentially means to provide support and prolong or preserve something. What we hope to preserve and support is life on planet earth. Thinking about this definition, we can view sustainability as an approach to life that causes the least possible harm to the natural world or living organisms.

Why is sustainability important?

The main goal of sustainability is to protect the planet so that future generations don't have to suffer. Ultimately, we only have finite resources on this planet, but we're currently not being considerate of that.

Our actions now shouldn't be at the expense of our fellow creatures or humans, so it's important to be aware of sustainability in our daily lives so we don't cause more irreparable damage to Earth.

What are the 3 pillars of sustainability?

The three pillars of sustainability are the three most fundamental aspects of sustainability, and they are environment, economy and society. All of them, directly and indirectly, impact each other – none of these pillars stand alone.

This is why it's important to consider everything when we're talking about sustainability, from materials and wastage to human treatment and working conditions.

What are some examples of sustainability?

There are so many different ways to be sustainable. One example is green energy – wind power, for example, is a free, natural, and infinite resource that doesn't cause harm to others, but does help to keep society functioning.

Another good example is creating, maintaining and looking after green spaces. This is because green spaces and plants improve air quality, protect water quality and reduce soil erosion, to name a few benefits.

How do you teach sustainability?

Sustainability should be taught to young children so that they embed healthy practices into their everyday lives without even thinking about it. For example, teaching them about composting, recycling, caring about animals and plants, and shopping second hand from a young age.

In later life, there are so many great resources, videos and sustainability courses that can teach people how to live more sustainably.

What are the benefits of sustainability?

The benefits of sustainability are countless, but to simplify this answer, we'll break it down into thinking about the three pillars of sustainability. Sustainable practice benefits the environment by conserving and looking after Earth's resources, preventing global warming and extreme weather, and protecting lives.

It benefits the economy by reducing wasted time, effort and power and finding a balance between growth and responsibility. It benefits society by building up communities and supporting those who are most vulnerable.

Is being sustainable difficult?

It's not always easy trying to make sustainable choices in a society where overconsumption and endless consumerism is encouraged, not only by the world's biggest corporations, but also by some governments.

However, more companies and governments are paying attention to sustainability than ever before, making it easier to make more informed and ethical choices. It may require a bit more research, but the benefits are endless.

Do you need to be rich to be sustainable?

You may have noticed that a lot of sustainable shops, whether they're selling clothes, food or skincare, are pretty expensive. This makes it appear quite costly to be sustainable, but it needn't be the case.

A lot of sustainable practices actually originated from a desire to save money; for example, using less energy, not buying unnecessary products and not wasting food. In addition, indigenous people are some of the most sustainable of all, despite having little money or material possessions.

Where can I learn more about sustainability?

Whether you're a new learner or someone looking to make your business more sustainable, FutureLearn offers flexible online sustainability courses that allow you to study the subject so that you can improve your understanding and potentially make important life changes.

What sustainability certifications or frameworks exist?

This depends greatly on the area you're interested in learning more about. You can find out about sustainable fashion certifications, sustainable food certifications, and more online. Certifications and frameworks help you to find companies that you trust, and you can then recommend to others who want to start their sustainability journey.

5.2 Questions on Living sustainably

How can you be more sustainable?

There are plenty of things that you can do to be more sustainable. What you're able to do will depend on things like where you live and how much time you have, but there are still options for everyone.

In addition, you don't have to be perfect in order to be sustainable. It's better to make a few changes and stick to them rather than not try at all because you're unable to do everything sustainably.

What are some examples of sustainable living?

As we've previously mentioned, sustainable living will look different for everyone. For some people, it might look like being completely self-sufficient – growing your own food, creating your own energy and building your own structures.

However, for others, sustainable living is merely being more aware of your choices, including how much you consume, where you buy from, and how you treat the Earth and other living creatures.

How can I be more sustainable at home?

There are so many simple ways you can be more sustainable at home. To start with, you can make sure you turn off any electric appliances or lights when you're not using them, reduce your water intake (take shorter showers), and recycle when you can. We'll go into more specifics in the questions and answers that follow.

How can I eat more sustainably?

In our blog Is eating meat bad for the environment? we investigate exactly how damaging the meat industry is. The main takeaway is that reducing your meat consumption can be a great way to have a more sustainable diet.

Other than that, you should think about where you get your food from, as small markets and butchers might be better than big supermarkets. In addition, try to buy items with as little single-use plastic packaging as possible.

How can I reduce my food waste?

There are so many ways you can reduce your food waste. The first thing is to keep track of the food you have in your fridge and cupboards. You can do this by making sure you keep older items near the front and making meal plans that use up all of your food items.

Then, you can compost any vegetable peels and odd bits, or use a food sharing app like Olio to share any food that is about to go off and you know you won't eat.

What is sustainable fashion?

Sustainable fashion can come in different forms. It's mostly sustainable to buy second-hand clothes from thrift stores, charity shops, and apps like Depop, so long as you don't then try to sell them for much more money afterwards.

It's also sustainable to look after the clothes you already own by caring for them properly and sewing or tailoring them when they break. Brand-wise, sustainable fashion is made with eco-friendly materials, designed and manufactured ethically, and ensures that everyone involved is paid fairly.

What's the most sustainable way to travel?

A huge amount of carbon emissions come from transportation, with a large majority of those emissions coming from road vehicles. Flying is also not great for the environment.

The most sustainable way to travel is either by foot or bike, but this is obviously not possible in many scenarios. Public transport like buses and trains are a good option, and if you do need a car for work, consider buying an electric or even hydrogen-powered car.

What is sustainable technology?

Sustainable technologies tend to use renewable energy, are made from materials that don't have a negative environmental impact, and are usually energy-efficient. Examples include electric vehicles, LED lights, carbon storage technologies and solar panels.

How can I bank sustainably?

You may not have considered banking on your sustainability journey before, but switching banks is actually one of the most impactful decisions you can make. This is because big banks are not only the cause of many financial crises, but they're also the biggest funders of fossil fuels.

Some examples of ethical banking alternatives include Triodos and Nationwide. App alternatives like Monzo and Starling are also better than the traditional banking giants.

How can I reduce my energy usage?

Transitioning to renewable energy is a beneficial step you can take, and organizations like Big Clean Switch facilitate this process with ease. If you are satisfied with your current energy provider, consider monitoring your energy consumption by installing a smart meter or participating in our Sustainable Energy Access for Communities course.

5.3 Questions on Sustainable development

What are the sustainable development goals?

The Sustainable Development Goals (SDGs) were established to outline a framework aimed at addressing the world's most pressing issues, with a target completion date of 2030. These goals encompass a range of critical areas, including the eradication of poverty, ensuring access to quality education, safeguarding life on land and in water, reducing inequalities, and combating climate change.

Who created the SDGs?

The Sustainable Development Goals were created in Rio de Janeiro in 2012 at the United Nations Conference on Sustainable Development. They served as a replacement for the Millennium Development Goals (MDGs) that were created in 2000 as part of a global effort to fight poverty and hunger.

How will the progress of the SDGs be monitored?

The SDG Tracker website provides comprehensive data on all indicators from the Our World in Data database, utilizing statistics sourced from the United Nations and other reputable international organizations. This platform serves as a free and open-access resource that tracks global advancements toward the Sustainable Development Goals (SDGs). By offering insights into their country's performance, citizens are empowered to hold their governments accountable for any shortcomings in implementing necessary changes.

What happens if we don't reach the SDGs?

There is a high chance that many of the SDGs won't be reached by 2030. While this won't all be related to COVID-19, the pandemic has certainly increased global inequity and made it harder to track data relating to the SDGs.

However, the goals are still important to have in place because they remind us what our hopes and aims for humanity are, and help us strive for something better.

How can design thinking help with sustainable development?

Design thinking serves as a vital approach in the context of sustainable development. Through the Designing for a Sustainable Future course offered by Samsung, participants can explore how technology companies leverage design thinking to address challenges and develop innovative solutions to global issues.

What is the difference between ESG and sustainability?

ESG is an acronym for environment, social, and governance, so we can see that the concept will have some overlap with sustainability. The main difference between the two is that ESG is specific and measurable, whereas sustainability is a broader and more vague term.

This means that companies can use ESG criteria to make tangible decisions and steps in their company to become greener, more ethical, and better governed.

What is sustainable construction?

Sustainability in construction plays a crucial role in fostering a more environmentally friendly future for upcoming generations. This approach emphasizes the use of renewable and recyclable materials while carefully considering energy consumption, resource utilization, and the ecological impact of the construction site.

If you are interested in expanding your knowledge of sustainable construction, you can develop your expertise and obtain construction certifications through our selection of online courses offered by leading universitie

What counts as a green building?

The World Green Building Council defines a green building as 'a building that, in its design, construction or operation, reduces or eliminates negative impacts, and can create positive impacts, on our climate and natural environment'. Features that can make a building green include the use of renewable energy, non-toxic materials and waste-reduction measures.

What does it mean to have a sustainable economy?

The economy is probably not something you normally associate with sustainability, but it's actually very important. Having a sustainable economy is about not favouring economic growth at the cost of social, environmental and cultural factors, and finding a good balance between growth and responsibility.

This isn't just for the benefit of the planet, but can also benefit big corporations by reducing wasted effort, time and money.

What is meant by the green economy?

A green economy takes this one step further. It's defined as low carbon, resource-efficient and socially inclusive by the UN Environment Programme, where public and private investments go into assets and infrastructure that work to reduce emissions, pollution and biodiversity loss, and to enhance energy and resource efficiency.

5.4 Questions on Careers in sustainability

What are some careers in sustainability?

If you're interested in sustainability, there are a bunch of careers you might be interested in, depending on your particular passion. You might want to work in renewable energy, wildlife conservation, animal welfare or green technology – but you can also work in an ordinary company and try to make sure they push for sustainable practices.

Are sustainability jobs in demand?

Sustainability is a fast-growing industry, so more and more jobs are becoming available. Interest from young people and students in sustainability is particularly driving the rise of sustainability careers. Some of the most in-demand jobs include clean transport manufacturers, urban growers and recycling plant operatives.

What are the benefits of working in sustainability?

Working in sustainability can be very rewarding because you know you're having a positive impact on the world. In addition, sustainable companies should pay and treat their workers well in order to reflect their values, and if they don't, it will be easier to hold them responsible. Finally, as sustainable initiatives are on the rise, you will have decent job security in this field.

How can you champion sustainability at work?

The best thing you can do is to talk to your colleagues and leadership team about sustainable practices. You don't need to preach to them, but by discussing options, pointing out the benefits of sustainability, and being sustainable yourself, you're likely to have a positive impact.

Why is corporate sustainability important?

Corporate sustainability is so important because it helps organisations strive for more sustainable practices and think about something other than profit. This is especially essential because just 100 companies are responsible for over 70% of emissions, so more responsibility needs to be taken by corporations than is currently happening.

How can I be more sustainable at work?

There are several things you can try to be more sustainable at work. Creating a green team that focuses on sustainable initiatives can be a good idea, or you could try to go paperless at work, ride your bike on your commute, or try to cut down on single-use plastics.

What does a sustainability officer do?

A sustainability officer (or CSO) has the role of analysing and predicting the stability and environmental impact of the company. They're in charge of setting policies, goals, and objectives that will help the company meet environmental policies as well as make a profit.

What does a sustainability consultant do?

A sustainability consultant has the job of working with different companies and businesses in order to help them become more responsible and conscious of sustainable and ethical practices. They evaluate the impact that the company is currently having by looking at things like their carbon footprint, and then they come up with tangible solutions.

How can I run a sustainable business?

Luckily for you, we have some great courses that'll help you get started with running a sustainable business. Try our Social Enterprise: Growing a Sustainable Business course or our Upcycling: From Green Ideas to Starting a Business course, both by Minca Ventures.

Can I study sustainability at university?

Yes, you can. They all vary a bit depending on which aspect of sustainability you want to eventually work in, but sustainability degrees are becoming increasingly common. From studying climate change to sustainable development or even sustainable fashion, there are plenty of options out there. One of our courses will make an excellent starting point.

5.5 Questions on Sustainable future

What is climate change?

Put simply, climate change is the long-term shift in weather patterns and global temperatures, often seen on a large scale. It's a natural part of life, and existed for centuries before humanity was born, but when we talk about climate change today, we're normally referring to the fast rise of temperatures that we've seen in the last 100 years as a result of human activity.

What are the biggest causes of climate change?

There are several main causes of climate change. The biggest one is the burning of fossil fuels, as this is the primary cause of global warming as a result of more greenhouse gases going into the atmosphere. The other big causes of climate change are deforestation, as this releases C02 and halts the release of oxygen, and many practices in the agricultural industry.

What is the impact of climate change?

The impact of climate change is extremely far-reaching, affecting our environment, fellow living creatures and society. Some of the biggest repercussions of climate change include extreme weather events like flooding and droughts that lead to forest fires; mass extinctions of animal and plant species; melting ice glaciers that lead to rising sea levels; and changing wildlife habitats that can negatively affect many ecosystems.

What are fossil fuels?

The three main fossil fuels are coal, crude oil and natural gas. They're named fossil fuels because they were formed from the fossils of plants and animals that lived millions of years ago, and their origins are the reason why they have a high carbon content. Fossil fuels are extracted using a range of methods, including mining, drilling, fracking and acidizing, all of which are extremely harmful.

What is renewable energy?

Renewable energy comes from energy sources that replenish themselves, and is, therefore, often naturally sustainable. For example, wind and sunlight are powered by nature itself, and so we can always rely on them for energy as long as the weather permits it. Renewable energy has been around for centuries in the form of waterwheels and windmills, but it was first used commercially in 1927.

What is my carbon footprint?

A carbon footprint can belong to an individual or company, and is essentially a measure of the total amount of greenhouse gasses released into the atmosphere as a result of your actions. It's usually measured in tonnes of CO2e. Reducing your carbon footprint is one tangible way you can try to make a difference and become more sustainable.

What does greenwashing mean?

You might have heard this term being thrown around lately, since greenwashing is becoming more popular in today's society. Greenwashing is when a company tries to market themselves as environmentally friendly or sustainable without actually working to reduce their environmental impact or become more ethical. Essentially, it's a deceitful advertising tactic used to appeal to those who want to shop sustainably.

How can we design for a sustainable future?

It will take a lot of work and commitment to ensure that we design for a sustainable future, and this requires us to educate ourselves. Luckily, we have plenty of courses created with that in mind.Try our Systems Thinking for Sustainability: Complex Systems Analysis Expert Track for a deep dive into complex systems analysis and sustainable design, or start off with our Tackling Environmental Challenges for a Sustainable Future course if you're new to this.

Can sustainable technology help save the world?

The role of technology is certainly a huge one when we contemplate saving the world from environmental, economic and social collapse. Technologies such as renewable energy and electric vehicles are undoubtedly game-changing, so we hope that even more solutions will arise in the near future.

However, we can't just rely on technology – human behaviour and simple actions will make all the difference.

What can we do as individuals to create a more sustainable future?

There are so many things we can do – in our section on living sustainably, we discussed that changing your food, fashion, transportation, banking and energy habits can make a real difference. In addition to this, you can push for legislation by going to peaceful protests, writing to your MP and local councillors, and advocating for sustainability on social media.

References

1. Scott J.J., de Bruin W.B., Rabinovich L., Frazzini A., de la Haye K. Understanding Food Insecurity in Los Angeles County during the COVID-19 Pandemic and Its Aftermath: A Qualitative Interview Study. Appetite. 2024;198:107323. doi: 10.1016/j.appet.2024.107323. [DOI] [PubMed] [Google Scholar]
2. UN World Population Prospects—Population Division—United Nations. [(accessed on 17 June 2024)]. Available online: https://population.un.org/wpp/
3. Manzoor S., Fayaz U., Dar A.H., Dash K.K., Shams R., Bashir I., Pandey V.K., Abdi G. Sustainable Development Goals through Reducing Food Loss and Food Waste: A Comprehensive Review. Future Foods. 2024;9:100362. doi: 10.1016/j.fufo.2024.100362. [DOI] [Google Scholar]

4. Munné-Bosch S., Bermejo N.F. Fruit Quality in Organic and Conventional Farming: Advantages and Limitations. Trends Plant Sci. 2024 doi: 10.1016/j.tplants.2024.01.011. [DOI] [PubMed] [Google Scholar]

5. Wittwer R.A., Klaus V.H., Oliveira E.M., Sun Q., Liu Y., Gilgen A.K., Buchmann N., van der Heijden M.G.A. Limited Capability of Organic Farming and Conservation Tillage to Enhance Agroecosystem Resilience to Severe Drought. Agric. Syst. 2023; 211:103721. doi: 10.1016/j.agsy.2023.103721. [DOI] [Google Scholar]

6. Gamage A., Gangahagedara R., Gamage J., Jayasinghe N., Kodikara N., Suraweera P., Merah O. Role of Organic Farming for Achieving Sustainability in Agriculture. Farming Syst. 2023; 1:100005. doi: 10.1016/j.farsys.2023.100005. [DOI] [Google Scholar]

7. Rifna E.J., Dwivedi M., Seth D., Pradhan R.C., Sarangi P.K., Tiwari B.K. Transforming the Potential of Renewable Food Waste Biomass towards Food Security and Supply Sustainability. Sustain. Chem. Pharm. 2024; 38:101515. doi: 10.1016/j.scp.2024.101515. [DOI] [Google Scholar]

8. Fami H.S., Aramyan L.H., Sijtsema S.J., Alambaigi A. The Relationship between Household Food Waste and Food Security in Tehran City: The Role of Urban Women in Household Management. Ind. Mark. Manag. 2021; 97:71–83. doi: 10.1016/j.indmarman.2021.06.016. [DOI] [Google Scholar]

9. Irani Z., Sharif A.M., Lee H., Aktas E., Topaloğlu Z., van't Wout T., Huda S. Managing Food Security through Food Waste and Loss: Small Data to Big Data. Comput. Oper. Res. 2018; 98:367–383. doi: 10.1016/j.cor.2017.10.007. [DOI] [Google Scholar]

10. Knorr D., Augustin M.A. From Kitchen Scraps to Delicacies to Food Waste. Sustain. Food Technol. 2024; 2:652–666. doi: 10.1039/D4FB00012A. [DOI] [Google Scholar]

11. Ashraf J., Javed A. Food Security and Environmental Degradation: Do Institutional Quality and Human Capital Make a Difference? J. Environ. Manag. 2023; 331:117330. doi: 10.1016/j.jenvman.2023.117330. [DOI] [PubMed] [Google Scholar]

12. Nguyen T.T., Grote U., Neubacher F., Rahut D.B., Do M.H., Paudel G.P. Security Risks from Climate Change and Environmental Degradation: Implications for Sustainable Land Use Transformation in the Global South. Curr. Opin. Environ. Sustain. 2023; 63:101322. doi: 10.1016/j.cosust.2023.101322. [DOI] [Google Scholar]

13. Newbery F., Qi A., Fitt B.D. Modelling Impacts of Climate Change on Arable Crop Diseases: Progress, Challenges and Applications. Curr. Opin. Plant Biol. 2016; 32:101–109. doi: 10.1016/j.pbi.2016.07.002. [DOI] [PubMed] [Google Scholar]

14. Launay M., Caubel J., Bourgeois G., Huard F., de Cortazar-Atauri I.G., Bancal M.-O., Brisson N. Climatic Indicators for Crop Infection Risk: Application to Climate Change Impacts on Five Major Foliar Fungal Diseases in Northern France. Agric. Ecosyst. Environ. 2014; 197:147–158. doi: 10.1016/j.agee.2014.07.020. [DOI] [Google Scholar]

15. Ghazoul J., Medina L. Deforestation and Land Clearing. In: Scheiner S.M., editor. Encyclopedia of Biodiversity. 3rd ed. Academic Press; Cambridge, MA, USA: 2024. pp. 605–616. [Google Scholar]

16. Cabral B.F., Yanai A.M., de Alencastro Graça P.M.L., Escada M.I.S., de Almeida C.M., Fearnside P.M. Amazon Deforestation: A Dangerous Future Indicated by Patterns and Trajectories in a Hotspot of Forest Destruction in Brazil. J. Environ. Manag. 2024; 354:120354. doi: 10.1016/j.jenvman.2024.120354. [DOI] [PubMed] [Google Scholar]

17. Gray E. Tropical Forests and Water Security in South America: Deforestation Trends, Drivers, and Solutions for Water Suppliers and Regulators. In: DellaSala D.A., Goldstein M.I., editors. Imperiled: The

Encyclopedia of Conservation. Elsevier; Amsterdam, The Netherlands: 2022. pp. 145–156. [Google Scholar]

18. Hung P.-Y. Food Nationalism beyond Tradition: Bubble Tea and the Politics of Cross-Border Mobility between Taiwan and Vietnam. Asian J. Soc. Sci. 2022; 50:222–228. doi: 10.1016/j.ajss.2022.08.008. [DOI] [Google Scholar]
19. Dupraz C.L., Postolle A. Food Sovereignty and Agricultural Trade Policy Commitments: How Much Leeway Do West African Nations Have? Food Policy. 2013; 38:115–125. doi: 10.1016/j.foodpol.2012.11.005. [DOI] [Google Scholar]
20. Nasir M.A., Nugroho A.D., Lakner Z. Impact of the Russian–Ukrainian Conflict on Global Food Crops. Foods. 2022; 11:2979. doi: 10.3390/foods11192979. [DOI] [PMC free article] [PubMed] [Google Scholar]
21. 10 Sustainable Solutions To Solve Global Warming - Sigma Earth
22. How gender inequality and climate change are interconnected | UN Women – Headquarters
23. notes on • Climate Justice and Equity - Search
24. 50 of the biggest sustainability questions answered - Future Learn
25. how melting ice affects people images - Search Images

Chapter-3

Waste Management

1.1 Introduction

1.1.1 Definition of Solid Waste

Solid waste encompasses a diverse array of discarded materials that individuals or organizations no longer find useful or necessary, with the intention of disposal. Typically, existing in solid or semi-solid states, these wastes arise from a multitude of human and industrial activities. The United States Environmental Protection Agency (EPA) defines solid waste as any refuse, including garbage, sludge from wastewater treatment facilities, and other discarded materials—whether solid, liquid, semi-solid, or contained gaseous—originating from industrial, commercial, mining, agricultural operations, and community active

The term encompasses a variety of materials, including:

i. Household garbage and rubbish

ii. Commercial and institutional wastes

iii. Industrial by-products

iv. Construction and demolition debris

v. Hazardous and non-hazardous materials

Solid waste management aims to handle these materials in ways that minimize their impact on the environment and human health, while also considering resource recovery, recycling, and sustainability.

1.1.2 Importance of Solid Waste Management

Effective management of solid waste is crucial for preserving environmental integrity, safeguarding public health, and promoting sustainable urban growth. With the rapid increase in population, industrial activities, and urban expansion worldwide, the challenges associated with waste generation are becoming more complex, underscoring the urgent need for efficient waste management strategies.

1.1.3 Environmental Protection

Improper disposal of solid waste can lead to soil, water, and air pollution. Leachate from landfills can contaminate groundwater, while the open burning of waste releases harmful pollutants into the atmosphere. Sound waste management practices reduce these environmental impacts by ensuring safe disposal, recycling, and treatment.

1.1.4 Public Health and Safety

Uncollected or improperly managed waste provides breeding grounds for pests such as mosquitoes, flies, and rodents, which can transmit diseases like malaria, dengue, and leptospirosis. Efficient collection and disposal prevent such risks and protect human health.

1.1.5 Resource Conservation

Recycling and composting help conserve natural resources by reducing the demand for raw materials. Materials like paper, metals, plastics, and organic waste can be recovered and reused, saving energy and reducing the environmental footprint of resource extraction and processing.

1.1.6 Economic Benefits

A well-organized waste management system creates employment opportunities in collection, sorting, recycling, and processing sectors. Additionally, revenue can be generated through the sale of recyclables, compost, or energy recovered from waste (e.g., waste-to-energy technologies).

1.1.7 Urban Aesthetics and Livability

Clean surroundings improve the quality of life for residents and promote tourism and investment. Proper waste management contributes to the overall cleanliness and orderliness of urban and rural areas.

1.1.8 Climate Change Mitigation

Waste that decomposes in landfills produces methane, a potent greenhouse gas. Reducing, recycling, and treating waste can significantly lower greenhouse gas emissions and contribute to climate change mitigation efforts.

1.1.9 Legal Compliance and Governance

Effective waste management helps municipalities and industries comply with environmental regulations and avoid legal penalties. It also promotes good governance and public trust in local institutions.

2.1 Overview of Challenges and Global Context

Solid waste management has emerged as a critical challenge in contemporary society, largely due to rapid urbanization, industrial expansion, and increasing consumer habits. The World Bank reports that over two billion tonnes of municipal solid waste are produced annually, with projections indicating this could rise to approximately 3.4 billion tonnes by 2050. This waste generation is not uniform; high-income nations produce the most waste per person, while low- and middle-income countries face more severe consequences from inadequate waste management systems, often hindered by insufficient infrastructure, policies, and funding. A significant obstacle in these regions is the inefficiency of waste collection systems, leading to a considerable amount of waste remaining uncollected, which exacerbates environmental issues and public health risks. In instances where collection does occur, the lack of source segregation hampers recycling and composting efforts, resulting in a reliance on landfills and open dumping. This practice not only contaminates soil and water resources but also contributes to the emission of harmful greenhouse gases such as methane.

Another critical issue is the absence of adequate treatment and disposal facilities. Many municipalities still rely on open dumping or unscientific landfills that lack proper lining and leachate management systems. Financial constraints further aggravate the problem, as many local authorities operate with limited budgets and insufficient technical expertise. Additionally, public participation is often low due to limited awareness, poor waste literacy, and weak engagement strategies. Without community involvement, even the best-designed systems may fail to achieve desired outcomes.

Regulatory and institutional challenges also pose significant obstacles. Many countries have outdated or poorly enforced regulations related to solid waste management. Inconsistent policies, overlapping responsibilities, and lack of coordination among government agencies reduce efficiency and accountability. However, in contrast, some high-income nations like Germany, Sweden, and Japan have demonstrated success by implementing advanced technologies, enforcing strict regulations, and promoting circular economy models that emphasize waste reduction, reuse, and recycling.

There is a growing global awareness of the importance of sustainable solid waste management practices. International initiatives, particularly the United Nations Sustainable Development Goals—especially SDG 11 and SDG 12—promote the development of sustainable urban environments and encourage responsible consumption and production. This has led to an increased emphasis on integrated solid waste management systems that address environmental, economic, and social factors. Such systems prioritize waste reduction at the source, efficient resource recovery, and the incorporation of digital technologies for enhanced monitoring and planning. As countries strive for sustainability and resilience, tackling the intricate challenges of solid waste management remains a critical focus.

3.1 Classification of Solid Waste

1. Municipal solid waste (MSW)
2. Industrial waste
3. Hazardous waste
4. Biomcdical waste
5. Agricultural waste
6. E-waste
7. Construction and Demolition Waste (C&D Waste)

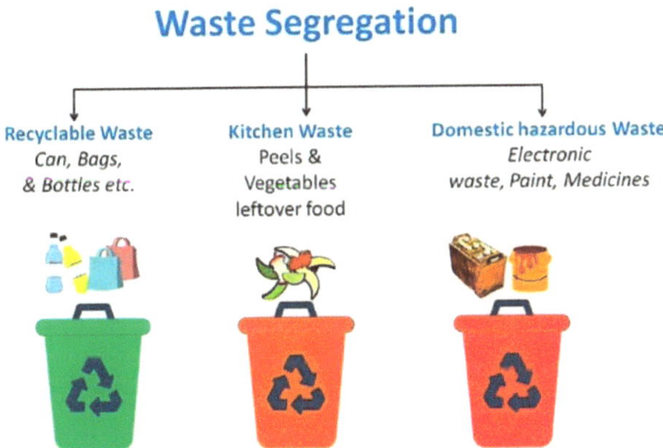

Figure 1 Classification of Solid Waste

3.1.1 Municipal Solid Waste (MSW)

Municipal Solid Waste (MSW), commonly known as trash or garbage, refers to the waste materials generated by households, commercial establishments, institutions, and municipal services. It includes a wide range of discarded items such as food scraps, paper, plastics, textiles, glass, metals, yard waste, and household hazardous materials shown in figure 1. MSW is a direct consequence of human activity and consumption patterns, and it presents significant challenges in terms of collection, transportation, treatment, and disposal, particularly in urban areas.

The composition of municipal solid waste varies greatly depending on factors such as geographic location, socioeconomic status, season, and urbanization level. For instance, developing countries typically have a higher proportion of organic waste due to greater reliance on fresh produce, while developed nations generate more packaging waste and disposables. Typically, MSW can be divided into biodegradable (e.g., food and garden waste), recyclable (e.g., paper, plastics, glass, metals), inert (e.g., construction debris), and residual waste that cannot be processed further.

Effective management of MSW is vital for maintaining public health, environmental quality, and the aesthetic appeal of cities. When not properly managed, MSW can become a breeding ground for pests and pathogens, contribute to the clogging of urban drainage systems, and lead to the contamination of air, water, and soil. Open dumping and unregulated burning of municipal waste are common practices in many developing countries, resulting in serious air pollution and greenhouse gas emissions, particularly methane from anaerobic decomposition of organic matter in landfills.

A sustainable MSW management system typically includes several stages: waste generation and storage, collection and transportation, sorting and processing, recycling and composting, energy recovery, and final disposal. Source segregation—separating waste at the point of generation into biodegradable, recyclable, and non-recyclable streams—is considered the foundation of an effective system. This facilitates recycling, reduces the burden on landfills, and enables better treatment of waste streams.

Recycling and composting are essential components of MSW management. Organic waste can be composted or converted to biogas through anaerobic digestion, reducing landfill volumes and producing valuable soil amendments or energy. Recyclable materials such as paper, plastic, and metals can be recovered and reprocessed, reducing the demand for virgin resources and energy use in manufacturing. However, this requires well-organized collection systems, public participation, and supporting infrastructure.

In recent years, cities around the world have adopted various approaches to improve MSW management. These include public-private partnerships, decentralized waste treatment facilities, pay-as-you-throw models to incentivize waste reduction, and the integration of informal waste pickers into formal systems. Digital technologies, such as GIS-based route optimization and smart bins, are also increasingly being used to enhance efficiency and accountability.

Ultimately, the goal of MSW management is not just to dispose of waste but to transition toward a circular economy where materials are reused and recycled, reducing the environmental impact and conserving resources. Achieving this requires coordinated efforts across all levels of society—governments, private sector, communities, and individuals.

3.1.2 Industrial Waste

Industrial waste refers to the unwanted or residual materials resulting from industrial processes and manufacturing activities. It encompasses a broad spectrum of substances, including solid, semi-solid, liquid, and gaseous waste generated by factories, processing plants, refineries, chemical plants, textile mills, mining operations, and other industrial establishments. These wastes may be hazardous or non-hazardous and can have serious environmental and health implications if not properly managed.

Industrial waste is typically categorized based on its composition and origin. Common types include chemical waste, metal and plastic scraps, ash, slag, sludge from wastewater treatment, packaging materials, and off-specification products. The nature and volume of waste generated depend on the type of industry. For example, the food processing industry produces organic waste, while the textile industry generates dye-containing effluents and fiber residues. Heavy industries such as steel and cement manufacturing produce large quantities of slag and dust, whereas electronics manufacturing may generate chemical solvents and e-waste.

The management of industrial waste poses significant challenges due to its complexity and potential toxicity. Many industrial processes use hazardous substances that, when released into the environment, can contaminate air, soil, and water. Toxic chemicals like heavy metals, volatile organic compounds (VOCs), and persistent organic pollutants (POPs) can have long-term effects on ecosystems and human health. Therefore, industrial waste is subject to stringent regulations, especially in developed countries, where industries must follow strict guidelines on waste storage, treatment, and disposal.

Industrial waste management typically involves a combination of reduction, reuse, recycling, and treatment before final disposal. The first step is waste minimization at the source through process optimization, cleaner production techniques, and material substitution. Reusing by-products within the same or other industries (industrial symbiosis) is another effective strategy. For instance, fly ash from thermal power plants can be used in the manufacture of cement or bricks. Recycling is feasible for many types of industrial waste, including scrap metal, solvents, and plastic components.

Treatment methods depend on the physical and chemical properties of the waste. Common techniques include incineration (for energy recovery and volume reduction), chemical neutralization (for acidic or alkaline waste), solidification/stabilization (for immobilizing hazardous components), and advanced methods like plasma arc treatment and pyrolysis. Liquid industrial waste is typically treated in effluent treatment plants (ETPs) or centralized wastewater treatment facilities before being discharged into water bodies or reused.

Despite regulatory frameworks and technological advancements, industrial waste remains a significant concern, particularly in developing nations where enforcement is weak, and informal practices are prevalent. Illegal dumping of untreated industrial waste in open lands, rivers, and drains is a common issue that leads to severe environmental degradation and public health hazards. Additionally, lack of awareness, inadequate monitoring systems, and limited access to treatment facilities further complicate the situation.

To address these challenges, a multi-pronged approach is required. This includes strengthening regulatory oversight, promoting cleaner production technologies, enhancing industry accountability through environmental audits and reporting, and encouraging investments in sustainable waste management infrastructure. Public-private partnerships, incentives for waste reduction, and international cooperation on best practices can also play a crucial role in improving industrial waste management systems.

3.1.3 Hazardous Waste

Hazardous waste refers to waste materials that pose a substantial or potential threat to public health or the environment due to their toxic, reactive, corrosive, or flammable properties. These wastes can exist in solid, liquid, or gaseous forms and are typically generated from industrial processes, healthcare facilities, laboratories, agricultural operations, and even household products. The improper handling and disposal of hazardous waste can lead to severe consequences, including soil and water contamination, air pollution, and long-term ecological damage.

The classification of hazardous waste is based on its characteristics. According to the Resource Conservation and Recovery Act (RCRA) of the United States Environmental Protection Agency (EPA), waste is considered hazardous if it exhibits one or more of the following traits: ignitability (easily combustible), corrosivity (ability to corrode metals or damage living tissues), reactivity (tendency to react violently with water or air), and toxicity (harmful when ingested, inhaled, or absorbed). Examples include used solvents, acids, heavy metals (like mercury, lead, and cadmium), pesticides, radioactive materials, and certain pharmaceutical wastes.

Hazardous waste is further categorized into several types, such as:

1. Industrial hazardous waste: Produced by chemical manufacturing, petroleum refining, and metal processing.
2. Biomedical hazardous waste: Includes sharps, infected materials, and toxic drugs.
3. Agricultural hazardous waste: Pesticides, herbicides, and veterinary pharmaceuticals.
4. Household hazardous waste: Batteries, paints, cleaning agents, and electronic devices.

The management of hazardous waste involves a highly regulated and systematic process to ensure environmental and human safety. The typical lifecycle includes generation, storage, transportation, treatment, and disposal. Each stage requires compliance with strict legal and safety standards, such as proper labeling, leak-proof containers, protective gear for handlers, and specialized transport vehicles. In most countries, industries are mandated to report hazardous waste generation and treatment practices to environmental regulatory authorities.

Treatment technologies for hazardous waste vary based on the type and nature of the waste. Incineration is commonly used to destroy organic hazardous materials and reduce volume, although it must be conducted in high-temperature, controlled facilities to avoid the release of toxic emissions. Stabilization and solidification techniques are employed to immobilize contaminants before land disposal. Chemical treatment, including neutralization and oxidation-reduction reactions, is applied to convert hazardous substances into less harmful forms. Some wastes require secure landfilling in specially engineered facilities with double liners, leachate collection systems, and groundwater monitoring.

Globally, hazardous waste generation is a serious concern. Developed countries generate large quantities of hazardous waste due to high industrial activity, but developing countries often bear the environmental burden when such waste is exported under the guise of recyclable materials. The Basel Convention, an international treaty, was established to control the transboundary movement of hazardous waste and ensure its environmentally sound management.

Challenges in hazardous waste management include lack of infrastructure, limited public awareness, and illegal dumping or mixing with non-hazardous waste. Many developing countries lack dedicated hazardous waste treatment facilities, leading to informal disposal practices that endanger communities and ecosystems.

Addressing these issues requires a combination of regulatory enforcement, technological advancement, capacity building, and public education. Promoting pollution prevention at the source, encouraging the use of safer alternatives, and integrating extended producer responsibility (EPR) are also crucial strategies to minimize hazardous waste generation and enhance sustainable management.

3.1.4 Biomedical Waste

Biomedical waste, also known as healthcare or medical waste, refers to any waste generated during the diagnosis, treatment, or immunization of humans or animals, as well as from research activities and the production or testing of biologicals. This category of waste is particularly concerning because it often contains infectious agents, toxic chemicals, sharps, and radioactive materials that pose significant risks to health and the environment if not properly managed.

Biomedical waste can be broadly classified into several categories, including infectious waste (such as discarded blood, tissues, and cultures), pathological waste (body parts and organs), sharps (needles, syringes, scalpel blades), pharmaceutical waste (expired or unused drugs), genotoxic waste (cytotoxic drugs used in cancer treatment), chemical waste (disinfectants, solvents), and radioactive waste (from diagnostic or therapeutic procedures involving radioactive materials). In healthcare settings, these wastes are generated in hospitals, clinics, laboratories, dental practices, veterinary institutions, and medical research facilities.

The improper handling and disposal of biomedical waste can lead to serious health hazards. Healthcare workers, waste handlers, patients, and the general public are at risk of infections such as HIV/AIDS, hepatitis B and C, and other blood-borne diseases through needlestick injuries or direct contact with contaminated materials. Environmental pollution is also a concern, as untreated biomedical waste can contaminate soil and water, and the open burning of such waste can release harmful dioxins and furans.

Effective biomedical waste management involves a well-organized system of segregation, collection, transportation, treatment, and disposal. The most crucial step is segregation at the point of generation, which ensures that infectious and hazardous waste is separated from general waste. Color-coded containers and bags are used for different categories of waste—for example, red for infectious waste, yellow for anatomical waste, blue for glassware, and white for sharps.

Treatment methods depend on the nature of the waste. Autoclaving (steam sterilization) is commonly used to disinfect infectious waste, while incineration is employed for pathological and pharmaceutical waste, especially when destruction is necessary. However, incineration must be carefully managed to prevent air pollution. Chemical disinfection, microwaving, and plasma pyrolysis are also used in certain contexts. Sharps are typically encapsulated or shredded after disinfection to prevent reuse. Final disposal may include landfilling of treated waste or ash in secured facilities.

In India and many other countries, biomedical waste management is governed by specific regulations. The Biomedical Waste Management Rules, 2016 (amended in 2018 and 2019) in India mandate all healthcare facilities to segregate, label, treat, and dispose of biomedical waste in accordance with national standards. Facilities must also maintain records and submit annual reports to the Pollution Control Boards.

Despite regulatory frameworks, challenges remain in many parts of the world, particularly in low-resource settings. These include inadequate infrastructure, lack of awareness and training among healthcare staff, poor monitoring and compliance, and informal waste handling practices. In some areas, biomedical waste is mixed with municipal waste or dumped in open spaces, leading to severe public health risks.

To ensure safe and sustainable biomedical waste management, there is a need for capacity building, strict enforcement of regulations, investment in modern treatment technologies, and widespread public education. Training healthcare workers, developing centralized treatment facilities, and promoting alternatives to incineration are essential steps toward improving the management of biomedical waste globally.

3.1.5 Agricultural Waste

Agricultural waste refers to the residues and by-products generated through various agricultural activities, including the cultivation of crops, rearing of livestock, aquaculture, and agro-industrial processing. This waste comprises both organic and inorganic materials, which, if not managed properly, can contribute to environmental pollution, greenhouse gas emissions, and degradation of natural resources. At the same time, agricultural waste holds significant potential for resource recovery, bioenergy production, and sustainable agriculture if appropriately utilized.

The primary types of agricultural waste include crop residues (such as straw, husks, leaves, and stalks), animal waste (manure, slurry, and carcasses), agro-industrial waste (such as bagasse, molasses, and oilseed cakes), and chemical waste (unused fertilizers, pesticides, and containers). Seasonal in nature, the quantity and composition of agricultural waste vary based on the type of crops grown, the scale of farming, climatic conditions, and farming practices.

One of the major concerns associated with agricultural waste is its impact on the environment. Open burning of crop residues, particularly in large agricultural economies like India and China, leads to air pollution, smog formation, and increased carbon emissions. The release of methane and nitrous oxide from improperly managed manure significantly contributes to climate change. Additionally, runoff from chemical fertilizers and pesticides contaminates water bodies, adversely affecting aquatic ecosystems and human health.

Traditionally, some agricultural waste has been recycled back into the farm ecosystem—for example, using crop residues as animal feed or mulch, and applying manure as organic fertilizer. However, with the intensification of agriculture and the introduction of synthetic agrochemicals, the volume of waste has increased, while traditional recycling methods have diminished. Furthermore, the lack of awareness and infrastructure for scientific waste management contributes to inefficiencies and environmental risks.

Effective agricultural waste management focuses on three key principles: reduction, reuse, and recycling. Waste minimization can be achieved through precision farming, efficient use of inputs, and integrated pest and nutrient management. Reuse and recycling involve converting waste into value-added products, such as compost, biochar, biogas, vermicompost, and organic fertilizers. Technologies like anaerobic digestion can convert animal waste into biogas for energy and digestate for soil amendment. Crop residues can also be used in the production of biofuels, paper, and biodegradable packaging.

Government policies and institutional frameworks play a crucial role in promoting agricultural waste management. Many countries have launched initiatives to encourage farmers to adopt sustainable waste practices through subsidies, training programs, and awareness campaigns. In India, for example, the National Policy for Management of Crop Residue aims to address stubble burning by promoting in-situ crop residue management and supporting machinery like happy seeders and rotavators.

However, significant challenges remain, particularly in rural and smallholder farming communities. These include a lack of technical knowledge, poor access to waste conversion technologies, limited market linkages for waste-based products, and inadequate financial support. Addressing these issues requires a multi-stakeholder approach involving farmers, extension services, government agencies, research institutions, and the private sector.

Agricultural waste, if managed sustainably, can contribute to a circular bioeconomy by transforming liabilities into assets. It supports soil health, renewable energy, and climate mitigation, while reducing the environmental footprint of farming. Promoting a shift toward sustainable agricultural waste management is therefore critical for the long-term health of ecosystems, food security, and rural livelihoods.

3.1.6 E-waste

E-waste, or electronic waste, encompasses discarded electrical and electronic devices that are no longer functional, such as computers, smartphones, televisions, and household appliances. The rapid advancement of technology, coupled with shorter product lifespans and increasing consumer demand, has led to a dramatic rise in global e-waste generation. The Global E-waste Monitor reported that around 53.6 million metric tons of e-waste were produced worldwide in 2019, with projections indicating an increase to 74.7 million tons by 2030. This escalating volume of e-waste presents serious environmental, health, and social issues.

The challenge of e-waste is compounded by the presence of valuable metals like gold and copper alongside hazardous materials such as lead and mercury. Improper disposal of e-waste can result in the contamination of soil, water, and air, posing significant health risks to both humans and wildlife. For example, heavy metals can leach into groundwater, leading to long-term environmental degradation and health hazards for communities dependent on these water sources. Furthermore, informal recycling practices, common in many developing nations, often involve unsafe methods such as open-air burning and hazardous dismantling, which expose workers and surrounding populations to toxic substances.

The lifecycle of electronic products typically follows a pattern of production, consumption, and disposal. However, in many cases, disposal happens prematurely due to planned obsolescence (where products are designed to become outdated or unrepairable) or due to consumer demand for newer technologies. As a result, the valuable materials in e-waste are often not recovered, and significant environmental and economic losses occur.

One of the key challenges in e-waste management is the lack of proper collection and recycling infrastructure. In many low- and middle-income countries, large amounts of e-waste are illegally exported from developed nations, creating a massive informal sector that is ill-equipped to handle the waste responsibly. Informal e-waste recyclers, often referred to as "e-waste pickers," manually extract valuable metals by burning, shredding, and chemically treating electronic devices. These methods release toxic fumes and lead to contamination of the surrounding environment. Furthermore, there is a lack of public awareness and incentives for responsible disposal and recycling of electronics, which contributes to improper disposal practices.

To address these challenges, governments and industries around the world have implemented various policies and strategies. The Basel Convention aims to minimize the transboundary movement of hazardous waste, including e-waste, and to ensure its environmentally sound management. Many countries have introduced Extended Producer Responsibility (EPR) programs, which hold manufacturers accountable for the disposal of their products at the end of their useful life. EPR programs require manufacturers to take back and recycle products or pay fees into a fund for recycling and safe disposal. In the European Union, the Waste Electrical and Electronic Equipment (WEEE) Directive sets high standards for the collection, recycling, and recovery of e-waste.

Technologies for the recycling of e-waste have improved in recent years, with advancements in mechanical processing, chemical recovery, and closed-loop recycling systems. Modern recycling techniques can extract metals with a higher degree of efficiency and minimize environmental harm. For instance, hydrometallurgical methods involve the use of aqueous solutions to selectively recover valuable metals from e-waste, reducing the need for energy-intensive smelting processes. Additionally, eco-design principles in electronics manufacturing, such as modularity, ease of disassembly, and the reduction of hazardous substances, can significantly extend product lifecycles and facilitate recycling.

The growing focus on the circular economy also offers a promising solution to the e-waste problem. In this model, products are designed for reuse, repair, and recycling, ensuring that materials are retained in the economy and not discarded as waste. This reduces the environmental impact of e-waste and provides economic opportunities by creating markets for refurbished electronics and recycled materials.

However, despite these efforts, e-waste remains one of the fastest-growing waste streams globally. To tackle the issue effectively, a multi-stakeholder approach is needed that combines better collection systems, stricter regulations, consumer awareness campaigns, and technological innovations in recycling. The transition toward a circular economy, along with global collaboration on e-waste management, is essential to mitigate the environmental and social impacts of this growing challenge.

3.1.7 Construction and Demolition Waste (C&D Waste)

Construction and demolition (C&D) waste refers to the debris generated during the construction, renovation, and demolition of buildings, roads, and other infrastructure projects. This category of waste includes a wide variety of materials, such as concrete, bricks, wood, metals, glass, plastics, gypsum, asphalt, insulation, and hazardous substances like asbestos. With rapid urbanization, industrialization, and large-scale infrastructure development, C&D waste is increasingly becoming a significant component of municipal waste streams worldwide. In many cities, it accounts for up to 30% of total waste generation, leading to significant environmental challenges.

The primary types of C&D waste include:

1. Concrete, cement, and mortar: These materials are often the most abundant and heavy waste produced during demolition.
2. Wood and timber: Typically found in older buildings, these materials are also generated during construction, renovation, and demolition.
3. Metals: Steel, iron, aluminium, and copper are commonly found in the structural components of buildings, such as beams, rebar, and pipes.
4. Glass: Used in windows, doors, and facades, glass waste can pose challenges in terms of safe handling and recycling.
5. Plastics and packaging materials: These are primarily generated during the construction phase as materials are transported and packaged.
6. Hazardous materials: These may include asbestos, lead-based paints, and chemicals used in construction that pose health risks if not properly managed.

C&D waste poses significant environmental and health risks if not properly managed. When sent to landfills or improperly disposed of, it can contribute to land degradation, water pollution, and air contamination. Concrete, for example, is highly alkaline and can leach harmful substances into the surrounding environment. The disposal of C&D waste in unauthorized areas can lead to the loss of valuable natural resources, as many materials, such as metals, wood, and concrete, can be recycled or reused.

Inadequate management of C&D waste also contributes to the depletion of natural resources. Raw materials such as sand, gravel, and timber are increasingly being extracted to meet the growing demand for construction materials, leading to environmental degradation and loss of biodiversity. The production of these materials also generates significant carbon emissions, further contributing to global climate change.

To address these challenges, the management of C&D waste requires a holistic and integrated approach, focusing on reduce, reuse, and recycle strategies. The reduce aspect emphasizes minimizing waste generation at the source, often through careful planning, efficient design, and the use of sustainable construction practices. By incorporating modular and prefabricated construction techniques, waste generation can be reduced, and more precise measurements of material requirements can be made.

Reuse of materials is an essential part of sustainable C&D waste management. Reusing building materials—such as bricks, wood, steel, and glass—helps reduce the need for new raw materials and lowers construction costs. Structures can be designed for deconstruction, which makes it easier to salvage and reuse materials at the end of their life cycle. For example, steel beams and window frames can be removed and reused in new buildings or infrastructure projects, while concrete can be crushed and used as aggregate in new construction.

Recycling of C&D waste is an increasingly viable and necessary practice. Many materials found in C&D waste, such as metals, plastics, and concrete, can be processed and converted into new products. Concrete can be crushed and reused as road base material, or even in the production of new concrete mixtures. Metals like steel and copper can be melted down and reused in manufacturing processes, while wood can be turned into composite materials or used for energy generation. However, to ensure that these materials are efficiently recycled, adequate infrastructure and collection systems must be in place.

One of the most effective ways to manage C&D waste is to establish construction and demolition waste recycling facilities that can sort, process, and recycle materials. These facilities can significantly reduce the amount of waste that ends up in landfills. In some countries, governments have introduced extended producer responsibility (EPR) schemes that require construction companies to manage the entire lifecycle of the materials they use, from production to disposal.

Regulations and incentives also play a crucial role in reducing C&D waste. Many countries have adopted laws and regulations that require the segregation, recycling, and proper disposal of C&D waste. For example, in the European Union, the Waste Framework Directive requires member states to minimize the environmental impact of C&D waste by promoting recycling and the reduction of landfilling. Additionally, the Construction Waste Management Plan (CWMP) is increasingly being mandated for large construction projects to ensure proper waste handling and disposal.

C&D waste management not only helps conserve resources and reduce environmental impact but also offers economic benefits. By recovering and recycling valuable materials, the construction industry can lower material costs and reduce dependence on raw material extraction shown in figure 2.

3.1.8 Sources of Solid Waste

1. Residential
2. Commercial
3. Institutional
4. Construction and demolition

Figure 2 Sources of Solid Waste

Residential Waste

Residential waste, also known as household waste, is generated from the day-to-day activities of individuals or families living in a particular residence. This waste includes a variety of materials, including food scraps, paper products, plastics, glass, metals, textiles, and other items used in the home. It also includes hazardous materials such as batteries, cleaning agents, and expired medications. The composition of residential waste can vary depending on household income, lifestyle, and consumption patterns.

A significant portion of residential waste is organic waste, including food scraps and yard trimmings, which, if improperly disposed of, can contribute to greenhouse gas emissions through decomposition in landfills. Packaging waste, such as plastic containers, cardboard, and bottles, makes up another large component of residential waste. While recycling programs have helped divert some of these materials, large quantities still end up in landfills. E-waste, including discarded electronics like old phones, computers, and televisions, is another growing component of residential waste. Proper management and recycling of residential waste, especially organics and recyclables, are critical in reducing the environmental impact and promoting a more sustainable waste management system.

Commercial Waste

Commercial waste refers to waste generated by businesses, including retail establishments, office buildings, restaurants, and other commercial enterprises. This waste is similar in composition to residential waste but typically consists of larger quantities and a wider variety of materials. It includes paper products (such as packaging, cardboard, and office paper), food waste, plastics, glass, metal cans, textiles, and electronic waste. The nature of commercial waste depends heavily on the type of business and the services or goods being provided.

For example, a restaurant or foodservice business generates large amounts of organic waste in the form of food scraps, food packaging, and disposable items like napkins and plastic utensils. Retail stores generate significant amounts of packaging waste, including plastic bags, cardboard boxes, and product packaging materials. E-waste from office electronics such as computers, printers, and telephones is also a growing concern in commercial waste management.

Commercial establishments often have the resources and infrastructure to separate recyclable materials, such as paper, glass, and plastics, but improper disposal practices and inadequate waste segregation can lead to significant volumes of waste being sent to landfills. Waste diversion programs, such as recycling and composting, are critical to reducing the environmental footprint of commercial waste.

Institutional Waste

Institutional waste is generated by large organizations and institutions, including schools, hospitals, universities, government buildings, and other public or private institutions. The composition of institutional waste is diverse and largely dependent on the type of institution and its operations. For example, hospitals and healthcare facilities produce biomedical waste, including syringes, bandages, surgical gloves, and contaminated medical materials. They also generate significant amounts of paper waste, including patient records, administrative documents, and medical reports. Food waste is another common component in institutions that provide cafeteria services.

Educational institutions such as schools and universities typically generate paper waste from books, notebooks, and office supplies, as well as food scraps from student cafeterias. Government buildings and offices also generate large volumes of paper waste, including reports, forms, and administrative documents.

Managing institutional waste requires a comprehensive strategy to segregate, recycle, and dispose of waste responsibly. For healthcare facilities, special attention must be paid to the segregation and safe disposal of hazardous and biomedical waste to prevent contamination and health risks. Educational and government institutions can adopt paper reduction strategies, promote recycling programs, and encourage sustainable practices within their organizations.

3.1.9 Construction and Demolition Waste

Construction and demolition (C&D) waste is generated during the construction, renovation, and demolition of buildings, roads, and other infrastructure. The types of materials found in C&D waste include concrete, bricks, wood, metals, glass, plastics, gypsum, asphalt, insulation materials, and hazardous substances like asbestos. C&D waste is typically large in volume and heavy in weight, and its composition varies depending on the type of project and the materials used.

One of the most significant components of C&D waste is concrete and masonry debris, which includes broken concrete, bricks, and mortar. Wood waste, including framing lumber, plywood, and timber from demolitions, is another common material in C&D waste. Metals, including steel rebar, copper wiring, and aluminum, are also frequently found in construction and demolition sites. Asbestos, often found in older buildings, is a hazardous material that requires specialized handling and disposal to avoid health risks, such as lung cancer and mesothelioma.

The management of C&D waste is particularly challenging due to the size and diversity of materials involved. While many of these materials can be recycled or reused, such as concrete (which can be crushed and used as aggregate for road construction) and metals, improper disposal can result in environmental harm, including soil contamination, water pollution, and habitat destruction. Efficient waste segregation, recycling, and proper disposal practices are essential to reducing the environmental footprint of C&D waste and minimizing the need for raw material extraction.

4.1 Properties of Solid Waste

Understanding the properties of solid waste is essential for designing and implementing efficient waste management systems. These properties influence the selection of collection methods, transportation systems, treatment processes, and final disposal techniques. The key properties of solid waste are discussed below:

1. Physical characteristics
2. Chemical composition
3. Biological properties

4.1.1 Physical Characteristics

The physical characteristics of solid waste refer to measurable properties such as density, moisture content, particle size, and texture. These characteristics are important for planning waste handling, processing, and disposal operations.

i. **Density:** This is the mass of waste per unit volume, usually expressed in kg/m^3. It varies significantly based on the type of waste and its composition. For example, organic waste tends to be denser than plastic or paper waste. Density affects the design of storage containers, transport vehicles, and landfill capacity.

ii. **Moisture Content:** Moisture content represents the amount of water in the waste and is typically expressed as a percentage of the total weight. High moisture content is common in food and garden waste. It significantly impacts the weight of waste, its decomposition rate, and the calorific value if waste is used for incineration or energy recovery.

iii. **Particle Size and Shape:** The size and shape of waste materials influence the efficiency of processing equipment such as shredders, compactors, and separators. Smaller particles decompose more quickly and are easier to handle in biological treatment systems like composting.

iv. **Colour and Odor:** These sensory characteristics can be indicative of the type and condition of waste. Organic waste often emits foul odors due to microbial activity and decomposition.

4.1.2 Chemical Composition

Chemical composition refers to the elemental makeup of the waste and is critical for determining the suitability of treatment methods like incineration, composting, and recycling.

i. **Organic and Inorganic Content:** Solid waste typically contains both organic (carbon-based) and inorganic materials. Organic components include food scraps, paper, and textiles, which are biodegradable and can be treated biologically. Inorganic materials include metals, glass, and plastics, which are more suitable for recycling or landfilling.

ii. **Elemental Composition:** This includes carbon (C), hydrogen (H), oxygen (O), nitrogen (N), sulfur (S), and ash content. The proportions of these elements affect energy recovery options. For instance, high carbon and hydrogen content increases the waste's calorific value, making it suitable for energy generation.

iii. **Calorific Value:** Also known as heating value, this indicates the amount of energy released during combustion. It is an important parameter for waste-to-energy technologies. Dry plastics and rubber have high calorific values, while organic waste with high moisture content has lower energy potential.

iv. **Presence of Hazardous Chemicals:** Some waste contains heavy metals (e.g., lead, mercury, cadmium), pesticides, and solvents. Their presence requires special handling and treatment to avoid environmental contamination and health hazards.

4.1.3 Biological Properties

Biological properties are primarily related to the organic and biodegradable fraction of solid waste and determine its behavior during biological treatment processes like composting and anaerobic digestion.

i. Biodegradability: This refers to the ability of organic waste to be broken down by microorganisms. Biodegradable materials, such as food and yard waste, can be transformed into compost or biogas through microbial action. The biodegradation rate depends on moisture, temperature, pH, and oxygen availability.

ii. Putrescibility: This is the tendency of organic waste to decompose and produce foul odors. High putrescibility indicates rapid decomposition, which can attract pests and contribute to air and water pollution if not managed properly.

iii. Pathogen Content: Waste from medical facilities, slaughterhouses, and households can contain disease-causing microorganisms. These pathogens pose serious health risks and must be treated through sterilization, incineration, or safe burial to prevent disease transmission.

iv. Compostability: Organic waste with high nutrient content and suitable carbon-to-nitrogen (C/N) ratio is ideal for composting. Effective composting depends on proper segregation and the absence of non-biodegradable or toxic substances.

5.1 Solid Waste Management Hierarchy

1. Waste prevention
2. Minimization
3. Reuse
4. Recycling
5. Energy recovery
6. Disposal (landfilling and incineration)

The Solid Waste Management Hierarchy represents a systematic approach to managing waste based on the principle of prioritizing the most environmentally beneficial practices. It is a pyramid-shaped framework, with the most preferred options at the top and the least desirable ones at the bottom. The hierarchy aims to reduce the quantity of waste generated, promote sustainable resource use, and minimize environmental impacts. The six levels of the hierarchy are explained below:

5.1.1 Waste Prevention (Source Reduction)

Waste prevention, also known as source reduction, is the highest and most preferred level in the waste management hierarchy. It focuses on eliminating waste before it is even created. This can be achieved by designing products and processes in a way that reduces material use and avoids unnecessary packaging or overproduction.

For example, manufacturers can reduce waste by using lightweight materials or redesigning products to use fewer components. Consumers can practice prevention by choosing durable, repairable products and avoiding disposable items. In industries, adopting cleaner production techniques and efficient inventory management can also help prevent waste generation.

The key benefit of waste prevention is that it conserves natural resources, reduces pollution, and decreases the need for waste treatment and disposal, making it both environmentally and economically advantageous.

5.1.2 Waste Minimization

Waste minimization goes a step further by reducing the quantity and toxicity of waste after it is generated but before it is discarded. It involves practices that improve the efficiency of resource use during production and consumption.

Examples of waste minimization include:

1. Substituting hazardous materials with safer alternatives
2. Optimizing production processes to reduce scrap and reject rates
3. Improving packaging design to use fewer materials

4. Encouraging the use of multi-functional and modular products

By minimizing waste, organizations and households can save costs on raw materials, reduce environmental liabilities, and comply with regulations more easily. It is especially relevant in industrial sectors where large volumes of process waste are generated.

5.1.3 Reuse

Reuse involves extending the life of products or materials by using them again for the same or different purpose, without significant processing. It is a practical and cost-effective way to reduce waste and conserve resources.

Common examples of reuse include:

1. Reusing glass bottles and jars
2. Donating clothes, furniture, and electronics for further use
3. Using containers and bags multiple times
4. Repurposing construction materials in other projects

Reuse preserves the embodied energy of products—the energy used in their production, transportation, and packaging—and helps avoid the environmental impact associated with recycling or disposal. Encouraging a reuse culture requires public awareness and the development of collection and redistribution systems.

5.1.4 Recycling

Recycling involves collecting, processing, and converting waste materials into new products. It is an essential component of sustainable waste management, as it reduces the need for virgin raw materials and lowers greenhouse gas emissions from manufacturing processes.

There are two main types of recycling:

1. Closed-loop recycling: Materials like aluminum cans and glass bottles are recycled into the same type of product.
2. Open-loop recycling: Waste materials are converted into different products, such as turning plastic waste into textiles or road-building materials.

Recycling is especially important for materials such as paper, metals, plastics, and glass. However, it requires proper segregation at the source, efficient collection systems, and well-developed recycling infrastructure. Contaminated or poorly sorted recyclables can lead to ineffective recycling or rejection of entire batches.

5.1.5 Energy Recovery

When waste cannot be reused or recycled, energy recovery offers a way to extract value by converting waste into usable forms of energy, such as electricity, heat, or fuel. Technologies used for energy recovery include:

i. Incineration with energy capture (waste-to-energy plants)
ii. Anaerobic digestion (biogas from organic waste)
iii. Gasification and pyrolysis (conversion of waste to synthetic gases or oils)

Energy recovery helps reduce the volume of waste that goes to landfills and provides an alternative to fossil fuels. However, it requires careful emission control to avoid air pollution and must not undermine efforts for recycling or prevention. Only non-recyclable and non-hazardous waste should be targeted for energy recovery.

5.1.6 Disposal (Landfilling and Incineration without Energy Recovery)

Disposal is the least preferred option in the hierarchy and should be considered only when no other options are feasible. It involves the final placement of waste materials in landfills or through incineration without energy recovery.

i. Landfilling is the most common disposal method, involving the burial of waste in designated sites. Modern sanitary landfills are designed with liners and leachate collection systems to prevent soil and groundwater contamination. However, landfills require large areas of land and generate methane, a potent greenhouse gas.

ii. Incineration without energy recovery is used to reduce the volume of waste but is less sustainable because it does not reclaim energy or materials. It can also emit harmful pollutants if not properly managed.

Disposal should be used as a last resort after all efforts at prevention, reuse, recycling, and energy recovery have been exhausted. Emphasizing disposal over higher-order strategies leads to resource depletion and long-term environmental damage.

6.1 Collection and Transportation

Effective collection and transportation of solid waste are critical components of an integrated solid waste management system. These processes ensure that waste is efficiently removed from its point of generation and transported to treatment, recycling, or disposal facilities in a manner that is hygienic, cost-effective, and environmentally sound. Poorly managed collection and transportation can lead to littering, pest infestations, foul odors, and public health risks. The major elements of this stage are discussed below:

i. Methods of collection

ii. Collection frequency

iii. Transfer stations and logistics

6.1.1 Methods of Collection

There are various methods of waste collection, which can be broadly categorized into manual and mechanized systems. The choice of method depends on factors such as the type of waste, density of the population, street layout, and available infrastructure.

i. Curb side Collection: This is the most common method, where residents place their waste bins or bags at the curb for scheduled collection. It may involve single or dual collection for segregated waste (e.g., wet and dry). Vehicles with compactors are typically used to collect and transport the waste.

ii. Backdoor Collection: In this method, collection personnel collect waste directly from the doorstep of each residence. Though more labor-intensive, this method is suitable for areas with narrow streets or multi-storied buildings.

iii. Community Bin Collection: Waste is deposited by residents in large community bins placed at designated locations. Collection vehicles empty these bins periodically. While cost-effective, this method requires regular maintenance to prevent overflows and odors.

iv. Block Collection: In this system, a waste collection vehicle parks at a central location, and residents bring their waste directly to the vehicle during scheduled hours. This method is often used in high-density or informal settlements.

v. Mechanized Collection: In urban areas, advanced waste collection systems may involve automated trucks equipped with lifting arms for handling large containers. These systems are efficient but require standardized bins and road access.

vi. Efficient waste collection relies heavily on proper segregation at the source, which facilitates recycling, composting, and reduces the burden on landfills.

6.1.2 Collection Frequency

The frequency of solid waste collection must be determined based on the type of waste, population density, and climate. Regular and timely collection prevents accumulation, minimizes odor, and deters vectors such as flies, rats, and stray animals.

i. Daily Collection: Recommended for organic and food waste in densely populated urban areas or in hot climates where decomposition is rapid. This helps control odor and bacterial growth.

ii. Bi-weekly or Weekly Collection: Acceptable for dry or non-biodegradable waste like plastics, paper, or recyclables, which do not decompose quickly.

iii. Seasonal Adjustments: Collection frequency may be increased during festive seasons, rainy periods, or events when higher waste generation is expected.

A well-planned collection schedule optimizes labor, fuel, and vehicle usage. It must be supported by community awareness to ensure that residents dispose of waste at appropriate times and locations.

6.1.3 Transfer Stations and Logistics

As cities grow, direct transportation of waste from the point of collection to disposal sites becomes inefficient, especially when the disposal site is far away. Transfer stations serve as intermediate facilities where collected waste is temporarily stored, consolidated, and then transferred to larger vehicles for transport to the final processing or disposal site.

Function of Transfer Stations:

a) Reduce hauling time and fuel costs by using larger vehicles for long-distance transport.

b) Enable segregation, compaction, or partial sorting of waste.

c) Serve as data collection points for waste quantity and composition.

i. Design and Features:

a) Should be located strategically to reduce travel time.

b) Must include infrastructure such as weighbridges, compactors, material recovery units, and ventilation systems.

c) Should be designed to minimize environmental impacts such as noise, odor, and leachate runoff.

ii. Transportation Logistics:

a) Route optimization is crucial for minimizing travel distances and time.

b) Vehicle selection must consider the type of waste, road conditions, and payload capacity.

c) GPS and digital monitoring systems are increasingly used to track collection vehicles, ensure timely service, and manage fleet efficiency.

d) Integrated logistics planning ensures cost-effective and sustainable waste transport while maintaining cleanliness and reducing greenhouse gas emissions.

7.1 Treatment and Disposal Methods

TheTreatmentanddisposalarethefinalstagesinsolidwastemanagement(figure3). They aim to reduce the volume, toxicity, and environmental impact of waste that cannot be reused or recycled. Proper selection and implementation of treatment and disposal methods ensure safe, hygienic, and sustainable waste management. The major methods are discussed below:

i. Composting

ii. Incineration

iii. Sanitary landfilling

iv. Anaerobic digestion

iv. Advanced technologies (e.g., plasma gasification)

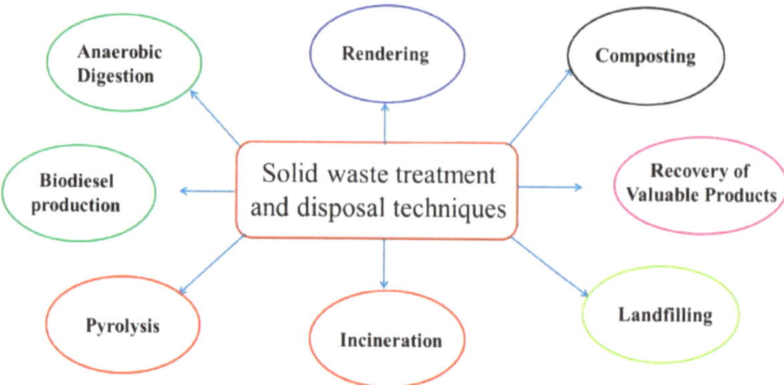

Figure3 Solid waste Treatment and Disposal Methods

7.1.1 Composting

Composting is a biological treatment method in which organic waste is decomposed by microorganisms under controlled aerobic (oxygen-rich) conditions to form a nutrient-rich soil amendment called compost.

- Process: Composting involves mixing biodegradable waste such as food scraps, garden waste, paper, and agricultural residues in the right carbon-to-nitrogen (C:N) ratio. Moisture content, aeration, and temperature are maintained to facilitate microbial activity. The process typically takes 4 to 8 weeks.
- Types:
  - Windrow Composting: Waste is arranged in long piles (windrows) and turned regularly.
  - In-vessel Composting: Composting occurs in enclosed containers with controlled parameters.
  - Vermicomposting: Earthworms are used to accelerate the decomposition process.

- Benefits: Reduces landfill burden, produces organic fertilizer, improves soil health, and lowers greenhouse gas emissions. It is ideal for household, agricultural, and municipal organic waste.
- Limitations: Requires source segregation, space, and management to prevent foul odor or pests.

7.1.2 Incineration

Incineration is a thermal treatment process where waste is combusted at high temperatures (800–1200°C) to reduce its volume and sometimes recover energy.

- Process: Waste is fed into a furnace or incinerator, where it burns in the presence of oxygen. The combustion reduces waste volume by 70–90%. Heat generated is often used to produce steam and electricity.
- Types:
    - Mass Burn Incinerators: Process unsegregated waste.
    - Refuse-Derived Fuel (RDF): Waste is pre-processed into uniform fuel before burning.
- Advantages: Rapid volume reduction, energy recovery, suitable for medical and hazardous waste.
- Disadvantages: High capital and operational costs, air pollution (dioxins, heavy metals), and public opposition due to environmental concerns. Advanced pollution control systems are necessary.

7.1.3 Sanitary Landfilling

Sanitary landfilling is the controlled disposal of waste in engineered landfills designed to minimize environmental and health risks.

- Design Features:

    - Liners: Prevent leachate from contaminating soil and groundwater.
    - Leachate Collection Systems: Collect and treat liquid runoff.
    - Gas Collection Systems: Capture methane gas for flaring or energy use.
    - Daily Cover: Waste is covered daily with soil to reduce odors, pests, and litter.
- Operation: Waste is compacted and layered, and the site is managed with regular monitoring.
- Pros: Can handle all types of non-hazardous waste, relatively low cost, long-term solution.
- Cons: Requires large land area, risk of leachate and gas emissions if mismanaged, not sustainable in the long term.
- Post-Closure: Landfills must be maintained for decades after closure to monitor and manage emissions and groundwater contamination.

7.1.4 Anaerobic Digestion

Anaerobic digestion (AD) is a biological treatment process where microorganisms decompose organic waste in the absence of oxygen, producing biogas (methane and carbon dioxide) and digestate.

- Process:
    - Waste is loaded into a digester.
    - In the absence of oxygen, microbes break down organic material.
    - Biogas is collected for use as fuel or electricity generation.
    - Digestate can be used as fertilizer.

- Suitable Waste: Food waste, manure, agricultural waste, and sewage sludge.
- Benefits: Produces renewable energy, reduces waste volume, lowers GHG emissions, and produces fertilizer.
- Challenges: Requires careful control of temperature, pH, and feedstock composition. Investment and operational costs can be high for small-scale operations.

7.1.5 Advanced Technologies

Advanced waste treatment technologies offer innovative approaches for dealing with complex and non-recyclable waste streams. One of the most promising among these is plasma gasification.

- Plasma Gasification:
  - Utilizes plasma torches to generate extremely high temperatures (up to 5000°C).
  - Waste is converted into syngas (mainly hydrogen and carbon monoxide) and a vitrified, inert slag.
  - Syngas can be used to generate electricity or as a chemical feedstock.
- Advantages:
  - Treats hazardous and toxic waste.
  - Reduces waste volume to nearly zero.
  - Produces clean energy with minimal emissions.
- Limitations:
  - Very high capital cost.
  - Requires technical expertise and stable waste input.

  - Still in early adoption phase globally.
- Other Technologies: Include pyrolysis, gasification (non-plasma), and mechanical biological treatment (MBT), each with specific applications and benefits.

8.1 Regulations and Policies

Effective solid waste management (SWM) depends not only on technology and infrastructure but also on robust regulations and policies that guide planning, implementation, and enforcement. Regulations provide the legal and institutional frameworks necessary to ensure safe, sustainable, and equitable waste management practices. At the national, international, and local levels, various agencies and bodies play vital roles in shaping these policies.

i. National and international policies

ii. Role of local government

iii. Legal frameworks

8.1.1 National and International Policies

National policies serve as the foundation for solid waste management by setting standards, defining responsibilities, and outlining strategic goals for waste reduction, treatment, and disposal. These policies are often developed by environmental ministries or central regulatory bodies and are enforced through subsidiary rules and guidelines.

i. In countries like India, the Solid Waste Management Rules, 2016, issued by the Ministry of Environment, Forest and Climate Change (MoEFCC), outline detailed responsibilities for waste segregation, processing, collection, and disposal. It also mandates waste generators to segregate waste at source and local bodies to establish waste processing facilities.

ii. In the United States, the Resource Conservation and Recovery Act (RCRA) governs solid and hazardous waste, emphasizing the reduction of waste generation and the protection of human health and the environment.

iii. European Union (EU) policies such as the Waste Framework Directive promote the waste hierarchy (prevention, reuse, recycling, recovery, and disposal) and support the circular economy concept.

At the international level, several agreements and initiatives address transboundary waste movement and environmental responsibility:

i. The Basel Convention controls the transboundary movements of hazardous wastes and their disposal.

ii. The Stockholm Convention deals with the elimination of persistent organic pollutants (POPs) in waste.

iii. The United Nations Sustainable Development Goals (SDG 11 and 12) encourage sustainable cities and responsible consumption and production, directly impacting waste management policy.

International cooperation helps countries adopt best practices, access funding for waste management infrastructure, and align with global sustainability goals.

8.1.2 Role of Local Government

Local governments are the frontline agencies responsible for the implementation and enforcement of solid waste management policies. Their roles include:

i. Collection and Transportation: Municipalities organize waste collection services, optimize routes, and ensure regular disposal.

ii. Infrastructure Development: Setting up waste segregation units, composting plants, recycling centers, landfills, and waste-to-energy plants.

iii. Monitoring and Compliance: Ensuring compliance with national rules regarding segregation at source, littering, illegal dumping, and management of different waste streams.

iv. Public Engagement: Running awareness campaigns, facilitating community participation, and incentivizing waste reduction and recycling practices.

v. Budgeting and Funding: Allocating municipal budgets for SWM services and exploring public-private partnerships to enhance efficiency.

vi. For local governments to succeed, they require adequate funding, technical expertise, political support, and community cooperation.

8.1.3 Legal Frameworks

Legal frameworks for solid waste management are structured to address waste generation, segregation, storage, collection, transportation, processing, and final disposal. These frameworks often include the following:

i. Environmental Protection Acts: Provide overarching guidelines for managing waste in a manner that protects the environment and public health.

ii. Waste Management Rules: Categorize and regulate different waste types (e.g., municipal solid waste, biomedical waste, e-waste, plastic waste, construction and demolition waste).

iii. Licensing and Permits: Waste processing units, transporters, and landfills must comply with licensing norms and operational standards.

iv. Penalties and Enforcement: Non-compliance with waste regulations results in fines, legal action, or closure of facilities. Regulatory bodies may conduct inspections and issue directives.

v. Extended Producer Responsibility (EPR): Legal provision mandating producers to take responsibility for post-consumer waste generated from their products, especially in the case of electronics, plastics, and packaging.

vi. A strong legal framework ensures accountability among all stakeholders—individuals, corporations, and government agencies. It also fosters transparency and encourages investment in cleaner technologies and infrastructure.

9.1 Challenges in Solid Waste Management

Despite advancements in technology and policy, solid waste management (SWM) continues to face several challenges across the globe. These challenges are particularly severe in developing countries and urban centers, where rapid population growth and industrialization outpace the capacity of existing waste infrastructure. Effective SWM requires a multi-dimensional approach that addresses technical, financial, institutional, and social factors. Below are three of the most critical challenges affecting solid waste management today:

i. Urbanization and population growth

ii. Financial and logistical constraints

iii. Public awareness and participation

9.1.1 Urbanization and Population Growth

One of the most significant challenges in modern waste management is the rapid pace of urbanization and population growth. As more people migrate to urban areas in search of better livelihoods, cities are expanding both in population density and spatial extent. This growth increases the volume and complexity of waste generated, including domestic waste, packaging materials, food waste, plastics, electronic waste, and construction debris.

Urban areas, especially in developing countries, often lack adequate planning and resources to manage this increase in waste. Informal settlements and slums may not have access to regular waste collection services, leading to open dumping and burning of waste. Moreover, urban sprawl makes the logistics of collection and transportation more complex and expensive.

Increased urban populations also generate more diverse and hazardous waste streams, such as industrial and electronic waste, which require specialized handling and treatment. This puts additional strain on municipal systems and highlights the urgent need for integrated urban planning that includes sustainable waste management strategies.

9.1.2 Financial and Logistical Constraints

Solid waste management is a capital-intensive service that involves expenditures for collection, transportation, processing, and disposal. Many municipalities, particularly in low- and middle-income countries, face financial limitations that hinder their ability to invest in modern equipment, skilled manpower, and infrastructure such as transfer stations, recycling centers, and sanitary landfills.

Revenue generation through user fees is often insufficient due to low collection rates or public resistance to pay for waste services. Inadequate budgeting and lack of access to external funding further exacerbate the problem. Even when funding is available, mismanagement and corruption can lead to substandard service delivery.

Logistical issues such as inefficient route planning, traffic congestion, lack of maintenance of vehicles, and absence of data-driven waste monitoring systems reduce the efficiency of collection and transport services. Additionally, landfill sites are often located far from urban centers, increasing fuel costs and time delays.

These financial and logistical bottlenecks not only reduce the effectiveness of waste management systems but also lead to environmental degradation and health risks.

9.1.3 Public Awareness and Participation

Public participation is essential for the success of any waste management initiative, yet it remains one of the weakest links in many communities. A lack of awareness about the importance of waste segregation, recycling, and responsible consumption results in poor cooperation from households, businesses, and institutions.

In many urban areas, waste is not segregated at the source, making it difficult to process or recycle efficiently. People often dispose of waste improperly due to ignorance, apathy, or inadequate infrastructure such as bins and collection points. This leads to littering, illegal dumping, and contamination of recyclable or compostable materials.

Changing public behavior requires continuous education and outreach programs. Schools, community groups, and media campaigns can play an important role in spreading awareness. Municipalities also need to enforce laws related to waste disposal and provide incentives for responsible behavior.

Furthermore, involving the informal sector—such as waste pickers and recyclers—in formal waste management systems can improve efficiency and offer livelihood opportunities, but this requires a shift in public and institutional attitudes.

10.1 Waste management interview questions

1. Question: What is waste management, and why is it important?

Answer: Waste management involves the collection, transportation, processing, and disposal of waste materials. It is crucial for environmental protection, public health, and resource conservation. Proper waste management helps minimize pollution and promotes sustainable living.

2. Question: How do you classify different types of waste?

Answer: Waste can be categorized as solid, liquid, or gaseous. Solid waste is further classified into biodegradable and non-biodegradable waste. Liquid waste includes sewage and industrial effluents, while gaseous waste comprises emissions and pollutants.

3. Question: How would you handle hazardous waste?

Answer: Hazardous waste must be treated with utmost care and disposed of following strict regulations. It involves specialized handling, storage, transportation, and treatment methods to prevent harm to humans and the environment.

4. Question: What are the primary challenges in waste management today?

Answer: The main challenges include improper waste segregation, inadequate infrastructure, lack of awareness, illegal dumping, and limited recycling facilities. Finding sustainable solutions to manage electronic waste and plastic pollution also remains a significant challenge.

5. Question: How can you promote waste reduction in a community?

Answer: Promote waste reduction through awareness campaigns, encourage recycling and composting, and support the implementation of waste-to-energy technologies. Partner with local authorities and community organizations to promote responsible waste management practices.

6. Question: What are the benefits of recycling?

Answer: Recycling conserves resources, reduces energy consumption, lowers greenhouse gas emissions, minimizes landfill space usage, and contributes to the circular economy by transforming waste into valuable resources.

7. Question: How can waste management contribute to a circular economy?

Answer: Waste management plays a vital role in the circular economy by promoting the reuse, recycling, and repurposing of materials, reducing waste generation, and encouraging sustainable production practices.

8. Question: What is the role of waste management in climate change mitigation?

Answer: Proper waste management reduces methane emissions from landfills, decreases the need for virgin materials, and minimizes energy consumption during production, all of which contribute to mitigating climate change.

9. Question: How do you stay updated on the latest waste management practices and technologies?

Answer: As a fresher, I will actively participate in workshops, webinars, and conferences, subscribe to industry journals and newsletters, and seek opportunities for professional development to stay informed about the latest advancements in waste management.

10. Question: What strategies can you suggest for promoting waste-to-energy technologies?

Answer: To promote waste-to-energy technologies, focus on educating the public about its benefits, engage with stakeholders, collaborate with the government and private sectors, and offer incentives for investing in such technologies.

11. Question: Describe a situation where you successfully managed waste in an eco-friendly manner.

Answer: As a fresher, I may not have direct experience, but I can provide a hypothetical situation where I organized a waste collection drive in my community and ensured proper segregation of waste for recycling and composting.

12. Question: How can waste management contribute to sustainable development?

Answer: By adopting sustainable waste management practices, we can reduce pollution, conserve resources, create green jobs, promote economic growth, and enhance the overall well-being of communities.

13. Question: What are your views on single-use plastics, and how can we address the plastic waste problem?

Answer: Single-use plastics are a significant environmental concern. To address this issue, we can advocate for plastic bans, promote alternatives, support recycling initiatives, and raise awareness about the importance of reducing plastic consumption.

14. Question: How do you handle conflicts with community members who resist waste management initiatives?

Answer: It's essential to engage in open dialogue, address their concerns, and highlight the long-term benefits of proper waste management. Building strong community partnerships and involving them in decision-making can help gain support for initiatives.

15. Question: What safety measures should be taken during waste collection and disposal?

Answer: Safety measures include using appropriate personal protective equipment, training workers in proper handling techniques, implementing secure waste storage and transportation, and adhering to local safety regulations.

16. Question: How can you encourage waste segregation at the source?

Answer: Educate residents about the benefits of waste segregation, provide clear instructions on proper segregation methods, and introduce incentive programs to encourage compliance.

17. Question: Describe the concept of Extended Producer Responsibility (EPR) in waste management.

Answer: EPR is a policy approach where producers are held responsible for the entire lifecycle of their products, including proper waste management and disposal. It encourages them to design products with recycling and environmental impact in mind.

18. Question: How can technology improve waste management practices?

Answer: Technology can enhance waste collection efficiency, optimize waste processing and disposal methods, enable better data management for decision-making, and support the development of innovative recycling technologies.

19. Question: In your opinion, what will be the future of waste management?

Answer: The future of waste management lies in a circular economy, advanced recycling technologies, increased emphasis on reducing waste generation, and sustainable waste-to-energy solutions to tackle the global waste crisis effectively.

20. Question: How do you plan to contribute to the field of waste management in the long run?

Answer: As a fresher, my goal is to continuously learn and grow in the waste management field. I aim to develop innovative and sustainable solutions, advocate for responsible waste management practices, and actively contribute to building a cleaner and greener future for all.

Armed with these essential waste management interview questions and answers, freshers can confidently approach their job interviews and demonstrate their knowledge, passion, and commitment to making a positive impact on the environment. Remember to emphasize the importance of sustainability, innovative solutions, and collaboration in waste management practices, and you'll be well on your way to securing your dream job in this vital and rewarding field. Good luck on your journey to becoming a waste management professional!

10.2 Waste management interview questions for experienced

Welcome to our comprehensive guide on waste management interview questions and answers for experienced professionals. As the world grapples with environmental challenges, waste management has become a critical aspect of sustainable development. In this blog, we will delve into a range of interview questions that experienced candidates may encounter, covering various waste management practices, innovative solutions, and strategies to tackle waste-related issues. Whether you're a seasoned waste management expert or a curious reader looking to enhance your knowledge, this blog is your go-to resource for acing your next waste management interview!

1. Question: How do you prioritize waste management initiatives in a corporate setting?

Answer: Prioritizing waste management initiatives in a corporate environment involves conducting waste audits to identify the most significant waste streams and potential areas for improvement. We focus on reducing, reusing, and recycling materials, implementing waste segregation programs, and encouraging employee engagement to foster a culture of sustainability.

2. Question: Can you describe a successful waste reduction project you led in the past?

Answer: In a previous role, I spearheaded a waste reduction project that involved replacing single-use plastic packaging with eco-friendly alternatives. By collaborating with suppliers and internal stakeholders, we achieved a 30% reduction in plastic waste and simultaneously lowered our carbon footprint.

3. Question: How do you ensure compliance with waste management regulations?

Answer: I stay updated on local, regional, and national waste management regulations and maintain clear communication with the relevant authorities. Additionally, I implement robust monitoring and reporting systems to track waste disposal practices and ensure compliance with all applicable laws and guidelines.

4. Question: What steps would you take to encourage waste segregation among employees?

Answer: To encourage waste segregation, I would conduct awareness campaigns, provide training sessions, and place informative posters near waste bins. Furthermore, I would reward employees who consistently adhere to waste segregation practices, fostering a sense of responsibility and ownership.

5. Question: How do you measure the effectiveness of a waste management program?

Answer: Key performance indicators (KPIs) such as waste diversion rates, reduction in landfill waste, and cost savings are essential metrics to measure the effectiveness of a waste management program. Regular audits and data analysis help track progress and identify areas for improvement.

6. Question: How do you handle hazardous waste disposal?

Answer: Handling hazardous waste requires strict adherence to safety protocols and relevant regulations. I ensure proper storage, labeling, and transportation of hazardous materials to licensed disposal facilities. Additionally, I work closely with qualified vendors to guarantee safe and compliant waste disposal.

7. Question: Have you implemented any innovative waste-to-energy solutions?

Answer: Yes, in my previous role, we established a waste-to-energy program where organic waste was processed through anaerobic digestion, generating biogas used to produce electricity. This initiative not only reduced waste but also provided a renewable energy source for the facility.

8. Question: How do you promote waste reduction among suppliers and vendors?

Answer: I engage suppliers and vendors in discussions about waste reduction initiatives, encouraging them to adopt sustainable packaging and materials. By including waste management criteria in procurement contracts and collaborating on sustainable practices, we collectively contribute to waste reduction efforts.

9. Question: Can you share an experience where you dealt with waste management challenges during a project?

Answer: During a construction project, we encountered excessive construction waste. To address this, I implemented on-site waste sorting and incentivized subcontractors to recycle materials properly, resulting in significant waste reduction and cost savings.

10. Question: How do you keep yourself updated on the latest waste management technologies and trends?

Answer: I attend industry conferences, participate in webinars, and read research papers and publications from waste management organizations. Additionally, I am an active member of waste management professional networks, which facilitates knowledge sharing and keeps me informed about the latest advancements.

11. Question: How do you integrate waste management into a company's sustainability strategy?

Answer: I align waste management goals with the company's overall sustainability strategy, emphasizing waste reduction as a key component. I collaborate with cross-functional teams to embed waste management practices into various processes, ensuring a holistic approach to sustainability.

12. Question: How would you handle a situation where an employee is resistant to waste management practices?

Answer: I would approach the employee with empathy and understanding, discussing the importance of waste management for the environment and the company's goals. By addressing their concerns and providing education, I aim to foster a sense of responsibility and encourage their active participation.

13. Question: Can you explain the concept of the circular economy in waste management?

Answer: The circular economy aims to minimize waste and maximize resource efficiency by designing products for reuse, recycling, and regeneration. In waste management, this involves promoting circular supply chains, product redesign, and embracing circular business models to reduce waste generation.

14. Question: How do you handle budget constraints when implementing waste management initiatives?

Answer: I prioritize waste management projects based on their potential impact and return on investment. By focusing on cost-effective solutions, collaborating with partners, and seeking external funding opportunities, I can make the most of the available budget.

15. Question: How do you engage the community in waste management efforts?

Answer: Community engagement is crucial for successful waste management. I organize workshops, town hall meetings, and educational programs to raise awareness and involve the community in waste reduction and recycling initiatives.

16. Question: What are your thoughts on the future of waste management and sustainability?

Answer: The future of waste management lies in embracing advanced technologies, circular economy principles, and collaborative efforts between industries, governments, and communities. Sustainability will continue to be a paramount focus, with waste reduction and responsible resource management playing pivotal roles.

17. Question: Can you describe a situation where you had to handle a waste-related emergency?

Answer: During a chemical spill incident, I promptly activated the emergency response plan, ensuring the area was cordoned off, and appropriate authorities were informed. I coordinated with emergency services and waste disposal experts to safely clean up and manage the hazardous waste.

18. Question: How do you motivate your team to stay committed to waste management goals?

Answer: I foster a culture of environmental responsibility by recognizing and celebrating team achievements, offering incentives for exceptional performance, and providing opportunities for professional development and growth in the waste management field. Continuous feedback and support keep the team motivated and focused on our waste management objectives.

We hope this compilation of waste management interview questions and answers has been insightful and valuable for all the experienced professionals out there seeking to make a positive impact on the environment. As our planet faces mounting waste-related challenges, the role of seasoned waste management experts becomes increasingly crucial.

Remember to stay updated on the latest industry trends, innovations, and best practices to remain at the forefront of sustainable waste management. Armed with the knowledge gained from this blog, we are confident that you will confidently navigate your way through any waste management interview and contribute significantly to creating a cleaner and greener future for generations to come. Good luck!

10.3 Waste management interview tips

1. Research the Company: Learn as much as you can about the waste management company you are interviewing with. Understand their services, projects, environmental initiatives, and any recent news or developments. Demonstrating this knowledge shows your genuine interest in the company and the industry.

2. Know the Industry: Familiarize yourself with waste management trends, challenges, and best practices. Understand the importance of waste reduction, recycling, and sustainability in today's world. Being well-informed about the industry will make you a more valuable candidate.

3. Highlight Relevant Experience: During the interview, be ready to discuss your previous experience and how it relates to waste management. Emphasize any work you've done in waste reduction, recycling, environmental compliance, or sustainability projects.

4. Showcase Problem-Solving Skills: Waste management often involves dealing with complex challenges. Be prepared to share examples of how you've approached problem-solving in your previous roles. Highlight instances where you identified waste-related issues and implemented effective solutions.

5. Emphasize Safety: Safety is a crucial aspect of waste management. If you have experience in adhering to safety protocols and standards, make sure to mention it. Show that you understand the importance of safety in waste handling and disposal.

6. Demonstrate Adaptability: Waste management can involve a variety of tasks and environments. Showcase your adaptability and willingness to work in different situations, whether it's handling hazardous waste, recycling materials, or working in outdoor settings.

7. Showcase Communication Skills: Effective communication is vital in the waste management industry, especially when collaborating with team members and conveying information about waste handling procedures. Be sure to express your communication skills during the interview.

8. Be Positive and Enthusiastic: Show enthusiasm for the waste management industry and the opportunity to work with the company. A positive attitude can go a long way in making a favorable impression on the interviewers.

9. Prepare Questions: At the end of the interview, the interviewer may ask if you have any questions. Be prepared with thoughtful questions about the company, the specific role, team dynamics, and opportunities for growth within the organization.

10. Dress Professionally: Dress appropriately for the interview, aiming for a professional and polished appearance. This shows respect for the opportunity and a willingness to make a good impression.
11. Practice Interview Questions: Review common interview questions and practice your responses. This will help you feel more confident and articulate during the actual interview.
12. Follow Up: After the interview, send a thank-you email or letter to express your gratitude for the opportunity. It's also an opportunity to reiterate your interest in the position and highlight any additional qualifications you may have thought of after the interview.

10.4 Waste management interview process

The waste management interview process typically involves several stages designed to assess your qualifications, skills, and fit for the position and the company. Here is a general outline of what you can expect during the waste management interview process:

1. Application Submission: The process usually begins with submitting your application, resume, and cover letter through the company's online application system or by email. Make sure your application highlights your relevant experience and skills in waste management or related fields.
2. Initial Screening: After reviewing applications, the company's HR team or recruiters will conduct an initial screening to shortlist candidates. This may involve reviewing resumes and cover letters, as well as conducting brief phone or video interviews to assess basic qualifications and interest in the position.
3. Technical Assessment: Depending on the role, you may be asked to undergo a technical assessment to evaluate your knowledge and skills related to waste management. This assessment could be in the form of a written test or an online evaluation.

4. First-round Interview: If you pass the initial screening and technical assessment, you will likely be invited for a first-round interview. This interview may be conducted in person, over the phone, or via video conferencing. During this interview, the interviewer will ask you about your background, experience, and interest in waste management. Be prepared to answer behavioral questions, technical questions, and questions related to your problem-solving abilities.

5. Second-round Interview: In some cases, there may be multiple rounds of interviews, especially for more senior or specialized positions. The second-round interview might involve meeting with different individuals or a panel of interviewers, including managers, team members, or executives. This stage may delve deeper into your technical expertise, teamwork skills, and cultural fit within the organization.

6. Site Visit or Job Simulation: Depending on the nature of the waste management role, you might be asked to participate in a site visit or job simulation. This could involve visiting a waste management facility, observing waste processing operations, or engaging in a simulated work scenario to assess your practical skills.

7. Reference Checks: Before making a final decision, the company may contact your references to verify your work history, skills, and performance. Ensure you provide accurate and up-to-date references who can speak positively about your qualifications.

8. Job Offer: If you successfully navigate through the interview stages and your references check out, you may receive a job offer. The offer will include details about the position, compensation, benefits, and other relevant information(figure4).

Throughout the waste management interview process, it's essential to be professional, enthusiastic, and knowledgeable about the waste management industry and the specific company you're interviewing with. Take the time to thoroughly prepare for each stage and tailor your responses to highlight how your skills and experience align with the organization's waste management goals and values. Good luck!

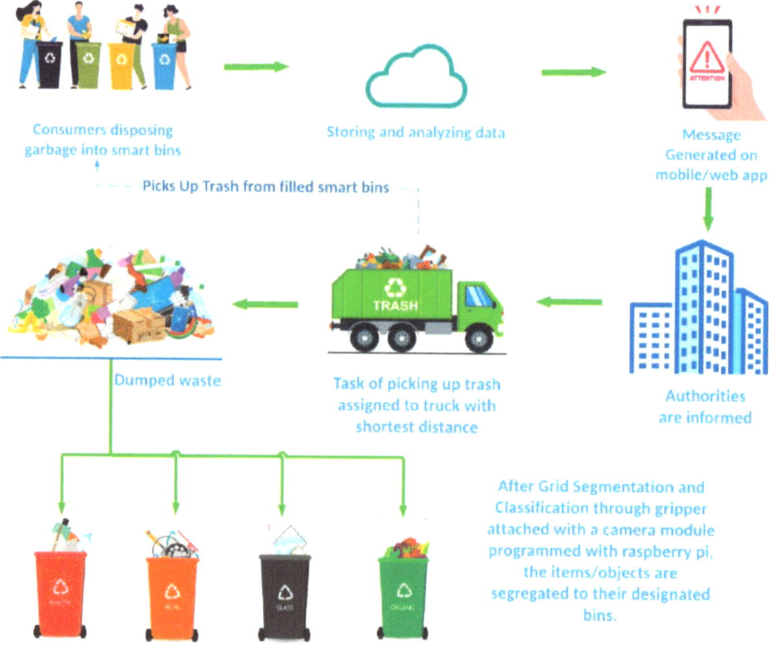

Figure 4 Waste management interview process

10.5 Multiple Choice Questions

1. Which of the following best defines solid waste?

   A) Liquid waste from industries

   B) Gaseous emissions from vehicles

   C) Unwanted solid materials discarded by society

   D) Rainwater runoff from roads

Answer: C

2. Which of the following waste types is commonly generated from households?

   A) Industrial waste

   B) Agricultural waste

   C) Municipal solid waste

   D) Hazardous waste

Answer: C

3. What is the primary cause of increased solid waste in urban areas?

   A) Decrease in industrial activity

   B) Increase in rainfall

   C) Urbanization and population growth

   D) Expansion of agricultural land

Answer: C

4. Which solid waste management method involves controlled biological decomposition of organic matter?

   A) Incineration

   B) Composting

   C) Landfilling

   D) Plasma gasification

Answer: B

5. What is the correct order of the Solid Waste Management Hierarchy?

   A) Disposal → Energy Recovery → Recycling → Reuse → Minimization

   B) Minimization → Reuse → Recycling → Energy Recovery → Disposal

C) Recycling → Disposal → Prevention → Reuse → Minimization

D) Energy Recovery → Prevention → Disposal → Recycling → Reuse

Answer: B

6. Which of the following is considered hazardous waste?

A) Kitchen scraps

B) Old clothes

C) Used batteries

D) Wooden furniture

Answer: C

7. Biomedical waste is primarily generated from:

A) Residential homes

B) Schools

C) Hospitals and clinics

D) Parks and recreation centers

Answer: C

8. Which of the following is an advanced technology used for waste treatment?

A) Open dumping

B) Incineration

C) Plasma gasification

D) Composting

Answer: C

9. Construction and demolition waste mainly consists of:

    A) Food waste

    B) Plastics and paper

    C) Concrete, bricks, and metals

    D) Medical syringes

Answer: C

10. Which sector is primarily responsible for policy implementation at the grassroots level?

    A) International NGOs

    B) Local Government

    C) National Parliament

    D) United Nations

Answer: B

11. Which of the following is NOT a biological property of solid waste?

    A) Biodegradability

    B) Pathogenic content

    C) Moisture content

    D) Presence of microbes

Answer: C

12. The Basel Convention deals with:

    A) Urban housing development

    B) Plastic recycling

    C) Transboundary movement of hazardous waste

    D) Employment in the informal sector

Answer: C

13. The key challenge in waste collection and transportation in developing countries is:

    A) Shortage of waste

    B) Lack of skilled drivers

    C) Financial and logistical constraints

    D) Excess vehicle availability

Answer: C

14. Which of the following methods is most suitable for treating food and garden waste?

    A) Landfilling

    B) Composting

    C) Incineration

    D) Plasma gasification

Answer: B

15. E-waste includes discarded:

    A) Newspapers and books

    B) Mobile phones and laptops

    C) Food packets

    D) Textile scraps

Answer: B

References

1. Tchobanoglous, G., Theisen, H., & Vigil, S. (1993). Integrated Solid Waste Management: Engineering Principles and Management Issues. McGraw-Hill.

2. Peavy, H. S., Rowe, D. R., & Tchobanoglous, G. (1985). Environmental Engineering. McGraw-Hill Education.

3. Cheremisinoff, N. P. (2003). Handbook of Solid Waste Management and Waste Minimization Technologies. Butterworth-Heinemann.

4. Kreith, F., & Tchobanoglous, G. (2002). Handbook of Solid Waste Management. McGraw-Hill.

5. Bilitewski, B., Hardtlein, M., & Marek, K. (1994). Waste Management. Springer.

Journal Articles

6. Hoornweg, D., & Bhada-Tata, P. (2012). What a Waste: A Global Review of Solid Waste Management. World Bank Urban Development Series.

7. Wilson, D. C., Velis, C., & Cheeseman, C. (2006). Role of informal sector recycling in waste management in developing countries. Habitat International, 30(4), 797–808.

8. Guerrero, L. A., Maas, G., & Hogland, W. (2013). Solid waste management challenges for cities in developing countrics. Waste Management, 33(1), 220–232.

9. Zaman, A. U. (2016). A comprehensive review of the development of zero waste management: Lessons learned and guidclines. Journal of Cleaner Production, 91, 12–25.

10. Aljaradin, M., & Persson, K. M. (2012). Environmental impact of municipal solid waste landfills in semi-arid climates–Case study–Jordan. The Open Waste Management Journal, 5(1), 28–39.

Government and Institutional Reports

11. CPCB (Central Pollution Control Board). (2016). Guidelines on Solid Waste Management. Ministry of Environment, Forest and Climate Change, Government of India.

12. Ministry of Housing and Urban Affairs (MoHUA). (2016). Solid Waste Management Rules, 2016. Government of India.

13. United Nations Environment Programme (UNEP). (2018). Africa Waste Management Outlook.

14. European Environment Agency (EEA). (2020). Waste Management in Europe: Current Status and Future Needs.

15. OECD. (2022). Improving Municipal Waste Management Services.

International Agreements and Frameworks

16. Basel Convention on the Control of Transboundary Movements of Hazardous Wastes and Their Disposal. (1989). United Nations.

17. Stockholm Convention on Persistent Organic Pollutants. (2001). United Nations Environment Programme.

18. UN-Habitat. (2010). Solid Waste Management in the World's Cities – Water and Sanitation in the World's Cities 2010. Earthscan.

Web Sources and Technical Papers

19. World Bank. (2018). What a Waste 2.0: A Global Snapshot of Solid Waste Management to 2050. https://datatopics.worldbank.org/what-a-waste

20. Ellen MacArthur Foundation. (2019). Circular Economy in Cities: Urban Planning for a Circular Economy.

21. Top 82 Waste Management Interview Questions 2024

Chapter-4

Disaster Management

1.1 What is a Disaster?

A disaster is characterized as a significant disruption, whether caused by natural events or human actions, that can occur over varying timeframes. Such events often result in severe human, material, economic, or environmental challenges that may exceed the coping capacity of the affected communities. In India, statistics indicate vulnerability to 30 distinct types of disasters, each posing a threat to economic, social, and human development. The impact of these disasters can be profound, leading to long-lasting repercussions on productivity and overall macroeconomic stability.

Disasters can be classified into the following categories:

- Geological Disaster: Landslides, earthquakes, volcanic eruptions, tornadoes
- Biological Disaster: Viral epidemics, pest attacks, cattle epidemic, and locust plagues
- Industrial Disaster: Chemical and industrial accidents, mine shaft fires, oil spills,
- Nuclear Disasters: Nuclear core meltdowns, radiation poisoning
- Man-made disasters: Urban and forest fires, oil spill, the collapse of huge building structures

1.1.1 What is Disaster Management?

In this section, we outline the concept of disaster management as defined by the Disaster Management Act of 2005. According to this legislation, disaster management encompasses a comprehensive approach that involves the planning, organization, coordination, and execution of essential measures aimed at effectively addressing disasters.

1. Prevention of threat of any disaster
2. Reduction of risk of any disaster or its consequences
3. Readiness to deal with any disaster
4. Promptness in dealing with a disaster
5. Assessing the severity of the effects of any disaster
6. Rescue and relief
7. Rehabilitation and Reconstruction

1.1.2 Agencies involved in Disaster Management

- The National Disaster Management Authority (NDMA) serves as the principal organization for disaster management in India, led by the Prime Minister. It oversees the National Disaster Response Force (NDRF), ensuring effective supervision, direction, and control in disaster response efforts.
- The National Executive Committee (NEC) comprises senior ministers from the Indian government, with the Union Home Secretary acting as Chairperson. This committee includes Secretaries from various ministries, such as Agriculture, Atomic Energy, Defence, and Environment, and is tasked with formulating the National Plan for Disaster Management in alignment with the National Policy on Disaster Management.
- At the state level, the State Disaster Management Authority (SDMA) is headed by the Chief Minister, who is supported by a State Executive

Committee (SEC). This committee plays a crucial role in assisting the SDMA with its disaster management responsibilities.

- The District Disaster Management Authority (DDMA) is led by the District Collector, Deputy Commissioner, or District Magistrate, depending on the circumstances, with elected local authority representatives serving as Co-Chairpersons. The DDMA is responsible for ensuring that all departments of the State Government at the district level, as well as local authorities, adhere to the guidelines established by the National Disaster Management Authority (NDMA) and the State Disaster Management Authority (SDMA).
- Local authorities encompass various entities, including Panchayati Raj Institutions (PRI), Municipalities, District and Cantonment Boards, and Town Planning Authorities, all of which oversee and manage civic services within their jurisdictions.

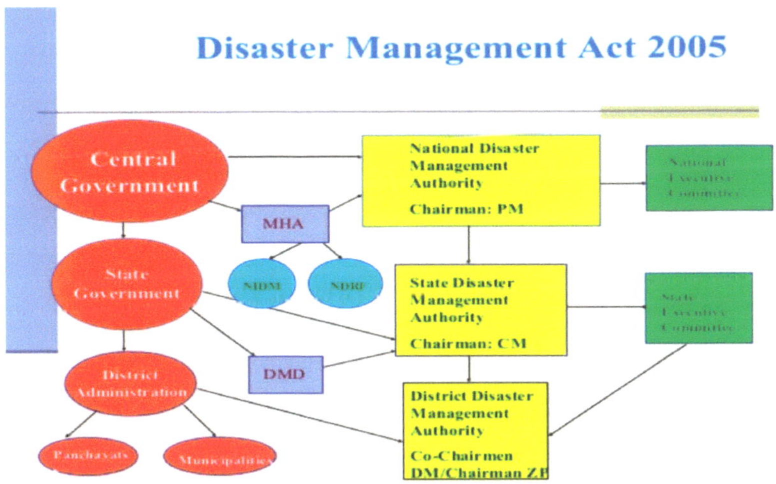

Figure1 Agencies involved in Disaster Management

2.1 Institutional Framework

In the Ministry of Health & Family Welfare (MoH&FW), public health needs to be accorded high priority with a separate Additional Directorate

General of Health and Sanitation (DGHS) for public health. In some states, there is a separate department of public health. States that do not have such arrangements will also have to take initiatives to establish such a department.

2.1.1 Operational Framework

At the national level, there is no policy on biological disasters. The existing contingency plan of MoH&FW is about 10 years old and needs extensive revision. All components related to public health, namely apex institutions, field epidemiology, surveillance, teaching, training, research, etc., need to be strengthened.

At the operational level, Command and Control (C&C) are identifiable clearly at the district level, where the district collector is vested with certain powers to requisition resources, notify a disease, inspect any premises, seek help from the Army, state or centre, enforce quarantine, etc. However, there is no concept of an incident command system wherein the entire action is brought under the ambit of an incident commander with support from the disciplines of logistics, finance, and technical teams, etc. There is an urgent need for establishing an incident command system in every district.

There is a shortage of medical and paramedical staff at the district and sub-district levels. There is also an acute shortage of public health specialists, epidemiologists, clinical microbiologists, and virologists.

Biosafety laboratories are required for the prompt diagnosis of the agents for the effective management of biological disasters. There is no BSL-4 laboratory in the human health sector. BSL- 3 laboratories are also limited. Major issues remain regarding biosecurity, the indigenous capability of preparing diagnostic reagents, and quality assurance.

Lack of an Integrated Ambulance Network (IAN). There is no ambulance system with advanced life-support facilities that are capable of working in biological disasters.

State-run hospitals have limited medical supplies. Even in normal situations, a patient has to buy medicines. There is a lack of stockpile of drugs, important vaccines like anthrax vaccine, PPE, or diagnostics for surge capacity. In a crisis, there is further incapacitation due to tedious procurement procedures.

2.2 National Disaster Response Force (NDRF):- The command and supervision of the NDRF would be under the Director-General of Civil Defence and National Disaster Response Force selected by the Central Government. Currently, the NDRF comprises of eight battalions who will be positioned at different locations as per the requirements.

Read about Crowd Disaster Management in the linked article.

3.1 Disaster Prevention and Mitigation

Proper planning and mitigation measures can play a leading role in risk-prone areas to minimize the worst effects of hazards such as earthquakes, floods, and cyclones. These are the key areas which should be addressed to achieve this objective:

- **Risk Assessment and Vulnerability Mapping:** Mapping and vulnerability analysis in a multi-risk structure will be conducted utilizing Geographic Information System (GIS) based databases like the National Database for Emergency Management (NDEM) and National Spatial Data Infrastructure (NSDI).
- **Increasing Trend of Disasters in Urban Areas:** - Steps to prevent unplanned urbanization must be undertaken, with the plan of action formulated being given the highest priority. State Governments/UTs concerned on the other hand focus on urban drainage systems with special attention on non-obstruction of natural drainage systems.
- **Critical Infrastructure:** - Critical infrastructure like roads, dams, bridges, irrigation canals, bridges, power stations, railway lines, delta water distribution networks, ports and rivers, and coastal

embankments should be continuously checked for safety standards concerning worldwide safety benchmarks and fortified if the current measures prove to be inadequate.

- **Environmentally Sustainable Development:** – Environmental considerations and developmental efforts, should be handled simultaneously for ensuring sustainability.
- **Climate Change Adaptation: -.** The challenges of the increase in the frequency and intensity of natural disasters like cyclones, floods, and droughts should be tackled in a sustained and effective manner with the promotion of strategies for climate change adaptation and disaster risk reduction.

4.1 Vulnerability Profile of India

India has been vulnerable to many natural as well as human-made disasters on account of its unique geo-climatic and socio-economic conditions. Out of 36 states and union territories in the country, 27 of them are disaster-prone.

Earthquake: According to the latest seismic zoning by the Bureau of Indian Standards (BIS), over 65 percent of the country is prone to earthquakes of intensity Modified Mercalli Intensity Scale (MSK) VII or more.The Himalayan region and the intra-plate faults are the regions of strong earthquakes in India.

Flood: Floods affect an average of 7.5 million hectares out of 40 million flood-prone areas in the country annually.Areas drained by the rivers of the Northern Plains are more prone to floods.

Cyclone and Tsunami: Of the 7,516 km long coastline, close to 5,700 km is prone to cyclones and tsunamis.

Drought: In India, around 68 percent of the agricultural land is prone to drought in varying degrees.Of the total region, 35 percent receives rainfall between 750 mm and 1125 mm, which is deemed drought-prone, while 33 percent receives rainfall less than 750 mm, which is termed chronically dry.

Landslides: Excluding snow-covered areas, approximately 12.6 percent of the country's geographical land area is prone to landslides.The North-western Himalayas are responsible for up to 66.5 percent of landslides, followed by the North-eastern Himalayas (18.8 percent) and the Western Ghats (14.7 percent).

5.1 Causes of Disasters in India

Disaster risks in India are compounded by its climatic conditions, changing demographic and socio-economic conditions, unplanned urbanization, environmental degradation, climate change, etc.

Geographical factors:

Tectonic plates: The geo-tectonic features of the Himalayan region and adjacent alluvial plains make the region susceptible to earthquakes, landslides, soil erosion, avalanches, etc. Location: India, being situated in the tropical region, is more prone to natural disasters such as cyclones and forest fires.

Rainfall patterns: India is mainly dependent on monsoon rain. The rainfall pattern is not uniform in India. In some places, like Assam and Bihar, there are frequent floods whereas; the western part of the country, including Rajasthan, Gujarat and some parts of Maharashtra are hit frequently by droughts.

Lack of infrastructure: The lack of adequate infrastructure and services, unsafe housing, and inadequate and poor health services can turn natural hazards into disasters. For example, Poor solid waste management may block stormwater and sewage networks, resulting in waterlogging and flooding. Further, rivers of Poor Urban Planning: Poor urban planning, inadequate drainage infrastructure, and rapid unregulated growth in cities has caused India's cities prone to urban flooding.

Deforestation and forest degradation: Deforestation leads to surface erosion, which is then linked to shallow, rapid landslides that are triggered by a single event of heavy rainfall. Deforestation also causes flooding in forest areas.

Climate change: Climate change has aggravated the risk of disasters across the world including India. Since warmer air can hold more water, it leads to heavier rainfall and extreme floods in some regions. Warmer temperatures enhance evaporation, which reduces surface water and dries out soils and vegetation and causes drought in some regions. Climate change increases sea surface temperatures, affecting tropical cyclones' formation and behaviour, leading to faster intensification, stronger wind speeds, and increased rainfall.

6.1 Types of Disasters

Primarily disasters are generally triggered by natural hazards human activity, or a combination of the two. The disasters are mainly classified into two types:

Natural disasters: The disasters that occur naturally are known as natural disasters, such as earthquakes, cyclones, tsunamis, heat waves, landslides, floods etc. These disasters occur both seasonally and without warning, subjecting the nation to frequent periods of insecurity, disruption, and economic loss as shown in figure 2.

Man-made disasters: Disasters that are caused due to human interruption are known as man-made disasters, such as industrial disasters, landslides, nuclear disasters etc.

Figure 2 Types of Disasters

6.1.1 Natural disasters

India is vulnerable to various natural disasters listed below.

Cyclone: Cyclones are created by atmospheric disturbances that surround a low-pressure area, characterised by rapid and frequently destructive air circulation. They are typically accompanied by strong storms and adverse weather.Every year, roughly five or six tropical cyclones emerge in the Bay of Bengal and the Arabian Sea. Out of these, two or three are catastrophic.

Drought: The primary cause of any drought is a deficiency of rainfall and in particular, the timing, distribution, and intensity of this deficiency to existing reserves. One-sixth area of the country is drought-prone. It resulted in crop failure, animal deaths, salinization of soil and groundwater decline, increased pollution of freshwater ecosystems, and regional extinction of animal species.

Tsunami: A tsunami is a series of water waves caused by the displacement of a large body of water, usually an ocean.The Geo-tectonic movements going on in the ocean floor make the coastal region prone to tsunamis. On-seismic events, such as a landslide or the impact of a meteor, may also lead to tsunamis. Japan is most prone to tsunamis. Tsunamis are anticipated to hit the Indian Ocean rim countries from two recognised tsunamigenic zones: the Andaman-Nicobar-Sumatra Island arc and the Makran subduction zone north of the Arabian Sea. The Tsunami of 26th December 2004 off the west coast of Sumatra, caused extensive damage to life and property in affected states in India.

Heatwaves: A Heat Wave is a period of exceptionally high temperatures that occurs throughout the summer season in North-Western India, usually between March and June, but can sometimes last until July. According to the IMD, heat waves should not be recognised until a station's highest temperature hits at least 40°C for plains and 30°C for hilly regions.Climate change is increasing the frequency of higher daily peak temperatures and longer, more intense heat waves around the world.

Landslide: Landslides constitute a major natural hazard in our country, which accounts for considerable loss of life and damage to communication routes, human settlements, agricultural fields, and forest lands. Landslides mostly affect the Himalayas and India's western ghats. The Himalayas are responsible for an estimated 30% of the world's landslides. In the Nilgiris alone, unprecedented rainfall in 1978 created almost one hundred landslides, causing significant damage. A valley in Nilgiris is called 'Avalanches Valley'.

Floods: India is highly vulnerable to floods. Floods in India are primarily caused by natural ecological systems like monsoons, silted river systems, and steep and highly erodible mountains, especially in the Himalayan ranges. Floods have also occurred in areas previously not considered flood-prone—for example, the 2022 Rajasthan floods.

Earthquakes: An earthquake is the release of energy that has been built up during the stress of increasing the formation of rock. This energy release takes the form of seismic waves that radiate in all directions from the place of movement. India has been categorised into four seismic zones based on the maximum magnitude of the projected earthquake. Zone V is the most active, covering all Northeast India, the northern part of Bihar, Uttarakhand, Himachal Pradesh, Jammu and Kashmir, Gujarat, and the Andaman and Nicobar Islands.

Cloudbursts: A cloudburst is a sudden spell of localised, heavy rainfall where the amount of rainfall over a particular area exceeds 100 millimetres in an hour. Cloudbursts occur when saturated clouds become unable to produce rain due to the upward passage of a highly warm circulation of air. Instead of dropping down, raindrops grow and are propelled up by the air current. They eventually get too heavy to hold and drop down, resulting in more rain than usual.

They often result in flash floods and have become increasingly common from May to September when the southwest monsoon season prevails in much of the country. In October 2023, a cloudburst above Lhonak Lake in North Sikkim caused a glacial lake outburst flood (GLOF) in the Teesta River that triggered a flood. GLOF (Glacial Lake Outburst Flood): A GLOF is a flood resulting from the sudden and rapid release of water from a glacial lake, often caused by the failure of a moraine dam or ice dam.

Climate change-induced glacial melt, seismic activity, and changes in glacial morphology are key risk factors for GLOFs. It can result in devastating downstream flooding, destruction of infrastructure, and loss of life. Rivers of the Himalayan regions, especially in Uttarakhand are most prone to the GLOF.

6.1.2 Man-made disasters

The human-induced factors can greatly aggravate the adverse impacts of a natural disaster. Human-induced factors like climate change, industrialization, and urbanization increase the frequency and intensity of extreme weather events, thereby escalating the potential for damage to life and property.

Figure3 Man-made disasters

Industrial and chemical disasters:

These are caused by chemical, mechanical, civil, electrical, or other factors like failures due to accident, negligence or incompetence in an industrial plant.

Chemical disasters are the occurrence of emission, fire or explosion involving one or more hazardous chemicals during industrial activity (handling), storage, or transportation or due to natural events leading to severe effects inside or outside the installation.

Both can result in extensive damage to the environment with considerable human and economic costs.

The Bhopal Gas Tragedy in 1984 was India's worst chemical disaster, resulting in over tens of thousands of deaths due to the accidental release of toxic gas Methyl Iso Cyanate.

Biological disasters: Biological disasters are caused by phenomena of organic origin or transmission by biological vectors, such as exposure to infectious microorganisms, toxins, and bioactive compounds.

7.1 Biological Disasters

7.1.1 Definition:

The devastating effects caused by an enormous spread of a certain kind of living organism that may spread disease, viruses, or an infestation of plant, animal, or insect life on an epidemic or pandemic level.

1. Epidemic Level – Indicates a disaster that affects many people in a given area or community.
2. Pandemic Level – Indicates a disaster that affects a much larger region, sometimes an entire continent or even the whole planet. For example, the recent H1N1 or Swine Flu pandemic.

To know more about Bio-Terrorism threat to India and India's Preparedness visit the linked article.

7.1.2 Biological Disasters - Important points to remember for UPSC

1. The nodal Ministry for handling epidemics – Ministry of Health and Family Welfare
    - Decision-making
    - Advisory body
    - Emergency medical relief providing

2. The primary responsibility of dealing with biological disasters is with the State Governments. (Reason – Health is a State Subject).
3. The nodal agency for investigating outbreaks – National Institute of Communicable Diseases (NICD)
4. Nodal ministry for Biological Warfare – Ministry of Home Affairs (Biological warfare is the use of biological agents as an act of war)

7.1.3 Biological Disasters - Classifications

The US Centres for Disease Control classifies biohazards into four biosafety levels as follows:

1. BSL-1: Bacteria and Viruses including Bacillus subtilis, some cell cultures, canine hepatitis, and non-infectious bacteria. Protection is only facial protection and gloves.
2. BSL-2: Bacteria and viruses that cause only mild disease to humans, or are difficult to contract via aerosol in a lab setting such as hepatitis A, B, C, mumps, measles, HIV, etc. Protection – use of autoclaves for sterilizing and biological safety cabinets.
3. BSL-3: Bacteria and viruses causing severe to fatal disease in humans. Example: West Nile virus, anthrax, MERS coronavirus. Protection – Stringent safety protocols such as the use of respirators to prevent airborne infection.
4. BSL-4: Potentially fatal (to human beings) viruses like Ebola virus, Marburg virus, Lassa fever virus, etc. Protection – use of a positive pressure personnel suit, with a segregated air supply.

8.1 Legislations for prevention of Biohazards in India

The following legislations have been enacted in India for the prevention of biohazards and implementation of protective, eradicative and containing measures when there is an outbreak:

1. The Water (Prevention and Control of Pollution) Act, 1974
2. The Air (Prevention and Control of Pollution) Act, 1981
3. The Environmental (Protection) Act, 1986 and the Rules (1986)
4. Disaster Management Act 2005, provides for the institutional and operational framework for disaster prevention, mitigation, response, preparedness, and recovery at all levels.

8.2 Prevention of Biological Hazards

The basic measure to prevent and control biohazards is the elimination of the source of contamination. Some of the prevention methods are as follows:

Preventive Measures for workers in the field (Medical)

1. Engineering controls – to help prevent the spread of such disasters including proper ventilation, installing negative pressure, and usage of UV lamps.
2. Personal hygiene – washing hands with liquid soap, proper care for clothes that have been exposed to a probably contaminated environment.
3. Personal protection equipment – masks, protective clothing, gloves, face shield, eye shield, shoe covers.
4. Sterilization – Using ultra heat or high pressure to eliminate bacteria or using biocide to kill microbes.
5. Respiratory protection – surgical masks, respirators, powered air-purifying respirators (PAPR), air-supplying respirators.

8.3 Prevention of Biological Hazards (Environmental Management)

Safe water supply, proper maintenance of sewage pipelines – to prevent waterborne diseases such as cholera, typhoid, hepatitis, dysentery, etc.

Awareness of personal hygiene and provision for washing, cleaning, bathing, avoiding overcrowding, etc.

Vector control:

Environmental engineering work and generic integrated vector control measures.

Water management, not permitting water to stagnate and collect and other methods to eliminate breeding places for vectors.

Regular spraying of insecticides, outdoor fogging, etc. for controlling vectors.

Controlling the population of rodents.

Post-disaster Epidemics Prevention

The risk of epidemics is increased after any biological disaster.

Integrated Disease Surveillance Systems (IDSS) monitors the sources, modes of diseases spreading, and investigates the epidemics.

8.4 Detection and Containment of Outbreaks

This consists of four steps as given under:

1. Recognizing and diagnosing by primary healthcare practitioners.
2. Communicating surveillance information to public health authorities.
3. Epidemiological analysis of surveillance data
4. Public health measures and delivering proper medical treatment.

Legal Framework for Biological Disasters

1. The Epidemic Diseases Act was enacted in the year 1897. (Read about RSTV's In-Depth Analysis on Epidemic Diseases Act 1897 in the linked article.)

2. This Act does not provide any power to the centre to intervene in biological emergencies.
3. It has to be substituted by an Act that takes care of the prevailing and foreseeable public health needs including emergencies such as BT attacks and the use of biological weapons by an adversary, cross-border issues, and international spread of diseases.
4. It should give enough powers to the central and state governments and local authorities to act with impunity, notify affected areas, restrict movement or quarantine the affected area, enter any premises to take samples of suspected materials, and seal them.
5. The Act should also establish controls over biological sample transfer, biosecurity and biosafety of materials/laboratories.

Biological disasters include epidemic disease outbreaks, plant or animal contagion, insect or animal plagues, and infestations. Biological disasters may be in the form of:

Epidemic– It affects a disproportionately large number of people in a population, group, or region at the same time; examples include cholera, plague, and acute encephalitis syndrome.

Pandemic– Pandemic is an epidemic that spreads across a large region, a continent, or even worldwide of existing, emerging, or reemerging diseases and pestilences, an example being Influenza H1N1 (Swine Flu) and COVID-19.

Urban floods: Rapid urbanisation and unplanned development along rivers and watercourses have led to increased runoff and urban floods due to the encroachment of sprawling habitations.

Flooding occurs quickly due to faster flow times (in minutes), hence called flash floods.

India has experienced an increasing number of urban flood disasters, with the most devastating ones being Srinagar in 2014 and Chennai in 2015.

8.0 Effects of the Disasters

Disasters, whether natural or man-made, have consequences for both humans and the environment. Different disasters bring with them different effects. It can be summarised below:

Loss of life and property: With a dense population living in a limited area, there is a tremendous loss of life and property. It also leads to the loss of agricultural and animal life.

Economic loss: Due to the destruction of public and private infrastructure, and loss of tourism the burden shifts on the public exchequer. According to global insurance giant Swiss Re, Natural disasters in India have caused uninsured economic damages of $32.94 billion (Rs 273,500 crore) during the five periods of 2018-22. Interruption of normal business: Disasters led to interruption in many services' transportation, communication, public gatherings, etc. Most of the companies in the country do not have a recovery plan in place for their vital IT infrastructure during times of both man-made and natural disasters.

Environmental losses: It includes natural resource depletion, reduced access to clean air, safe water, diminished biodiversity, soil degradation, etc.For example, landslides can wipe out large tracts of forest, destroy wildlife habitat, and remove productive soils from slopes.

Environmental loss due to forest fires such as the catastrophe forest fire in Australia 2019-20, frequent forest fires in Uttarakhand, and the Amazon forest fire are examples of environmental loss.Tropical cyclones significantly impact coastal communities, particularly coral reefs, affecting coral cover, species diversity, and reef productivity.

Health effects: Man-made disasters like nuclear accidents, oil spills, etc. significantly impact health. It may cause neurological disorders, acute radiation syndrome, etc.A major disaster in a country like India may induce after-effects like poverty, hunger, and malnutrition.... Read more at: https://vajiramandravi.com/upsc-exam/disasters/

9.1 What Is a Tsunami? Definition and Explanation

A tsunami is a series of enormous ocean waves that results from the rapid displacement of a large volume of water. The waves often rise to a height of over 30 meters (100 feet). Unlike typical ocean waves, which are caused by the wind, tsunamis are primarily the result of geological activities shown in figure 4.

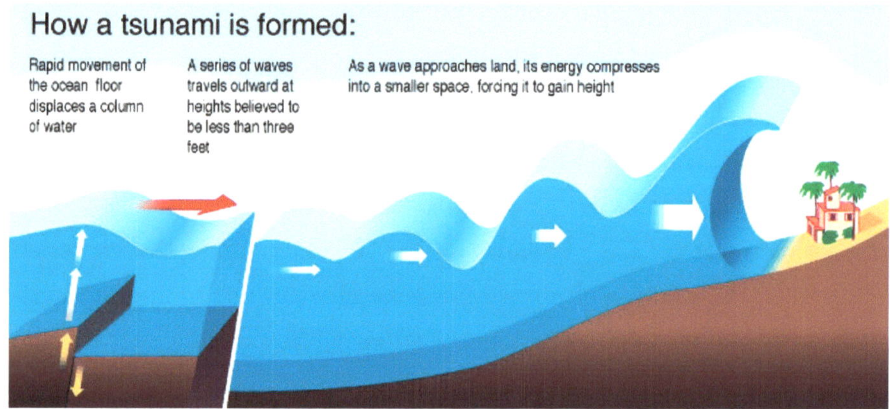

Figure 4 Tsunami wave formation

Word Origin and Comparison with Other Terms

The word 'tsunami' is of Japanese origin, where 'tsu' means harbor and 'nami' means wave, essentially translating to "harbor wave." This term is favored over alternatives like "tidal wave" or "seismic sea wave" because it captures the essence of the phenomenon more accurately.

- Tidal Wave: Tsunamis are not influenced by the tides, so the term "tidal wave" is misleading.
- Seismic Sea Wave: This term is closer to describing a tsunami but is somewhat restrictive, as seismic activity is just one of the causes.

Causes of Tsunamis

There are multiple causes for tsunamis, including:

1. Underwater Earthquakes: Undersea earthquakes are the most common cause of tsunamis, where tectonic plates shift suddenly. Aftershocks may generate additional waves.
2. Volcanic Eruptions: Explosive eruptions or the collapse of volcanic islands displace water, sometimes triggering a tsunami.
3. Landslides: Some tsunamis results from either underwater landslides or from a land mass sliding into the ocean. An ice mass breaking off and falling into the ocean is another potential trigger.
4. Meteorite Impacts: Although rare, a large enough meteorite impact on an ocean can generate a tsunami.
5. Human Events: A tectonic weapon has the potential for inducing a tsunami. Most explosions do not generate large waves, but the 1917 Halifax Explosion produced an 18-meter-high tsunami in the harbour.

Approximately 80% of tsunamis occur in the Pacific Ocean, but they can happen in any large body of water, including lakes. Shoreline topography is important, too. For example, Japan has experiences over a hundred tsunamis throughout history, while nearby Taiwan has only recorded two.

9.1.1 How a Tsunami Works

A tsunami starts with an event that displaces a large volume of water. The resulting waves spread outward radially, much like the pattern you see when you drop a rock into a pool. These wave move more quickly than wind waves and gain height when they reach shallow water. Unlike normal waves, tsunami waves rarely break. Instead, a tsunami appears as a wall of water or tidal bore.

1. Initiation: Geological activity displaces a large volume of water.
2. Propagation: The waves move outward in all directions from the point of origin.

3. Amplification: As the tsunami approaches shallower waters, it gains height.

4. Impact: The waves reach the shore, often with little warning, causing destruction.

A tsunami is a set of waves and not a single wave. It may feature multiple waves that arrive over a period of hours. The first wave is not always the highest one.

9.1.2 Tsunami Characteristics

The waves from a tsunami differ from ordinary waves:

1. **Long Wavelengths**: Unlike regular waves, tsunamis have wavelengths that can extend up to 200 miles. In other words, the distance from the trough of one wave to the next might be miles or kilometers, rather than the typical 60–150 m (200–490 ft) wavelength of wind-caused waves.

2. **High Speed**: They travel at speeds up to 500-800 km/h (310-500 mph). So, time is a critical factor in reducing the impact of the waves.

3. **Increase in Height**: Tsunamis are often barely noticeable in deep water but increase dramatically in height as they approach shallower waters. So, a ship in deep water might be unaffected by a tsunami that causes devastation on shore.

9.1.3. Recognizing a Tsunami

How do you know when a tsunami is coming? Warning systems are the best protection, but watching the water and maybe surrounding wildlife also helps.

Drawback

Before a tsunami hits, there is often a noticeable retreat of water from the shore, known as 'drawback.' This phenomenon serves as a natural warning signal. If you see the ocean retreating, head for high ground.

Warning Systems

Sophisticated early-warning systems, involving seismic sensors and ocean buoys, provide some advance notice. The notice ranges from minutes to hours, depending on the distance from the point of origin.

Animal Behavior

While not scientifically confirmed, there are numerous reports of animals acting unusually before tsunamis, possibly due to their sensitivity to vibrations or sounds humans can't detect.

Time to Safety The time to reach safety varies significantly, depending on how close the tsunami source is to the coastline. In some cases, people have just minutes.

Magnitude Scales Two of the most common tsunami magnitude scales are the Imamura-Iida Intensity Scale and the Sieberg-Abraseys Scale.

- Imamura-Iida Intensity Scale: This scale measures height and distance travelled.
- Sieberg-Ambraseys Scale: This scale measures effects on both humans and landscapes.

9.1.4 Mitigating Future Damage

Scientists and policymakers are taking a multi-tiered approach to minimizing the impact of future tsunamis. While the events are not preventable, improving warning systems and public education and building structures to withstand the waves reduces the damage and loss of life.

1. Improved Warning Systems: This include increasing the network of seismic and oceanographic sensors and establishing sirens and emergency evacuation routes.
2. Engineered Structures: Building seawalls and breakwaters, as well as engineering buildings reduces the impact of the waves..

3. Community Preparedness: Education and drills reduces the time it takes for people to take action and reach safety.

9.1.5 Major Historical Tsunamis

Here are 10 historically important tsunamis:

1. **Indian Ocean, 2004**: One of the deadliest natural disasters in recorded history, this tsunami was triggered by a massive undersea earthquake off the coast of Sumatra, Indonesia. It resulted in over 230,000 deaths across 14 countries, including Thailand, Sri Lanka, and India.
2. **Tohoku, Japan, 2011**: Triggered by a 9.0-magnitude earthquake, this tsunami led to the Fukushima nuclear disaster. Nearly 16,000 people were killed, and the event had extensive economic repercussions.
3. **Lituya Bay, Alaska, 1958**: The highest tsunami wave ever recorded occurred in Lituya Bay, Alaska, with a wave reaching 1,720 feet. Triggered by a landslide, it had a relatively lower human toll but showcased the incredible power of tsunamis.
4. **Great Lisbon Earthquake and Tsunami, 1755**: Occurring on All Saints' Day, this event devastated Lisbon, Portugal, and affected much of Europe and North Africa. The tsunami wave traveled as far as the Caribbean.
5. **Krakatoa, Indonesia, 1883**: The eruption of the Krakatoa volcano resulted in a tsunami with waves as high as 135 feet. The event was so powerful that it was heard 3,000 miles away, and it killed approximately 36,000 people.
6. **Messina, Italy, 1908**: Triggered by an earthquake in the Strait of Messina, this tsunami killed an estimated 80,000 people in the cities of Messina and Reggio Calabria.

7. **Nankaido, Japan, 1707**: This is one of the earliest well-documented tsunamis. It resulted from a massive earthquake and caused significant loss of life and property in Japan.

8. **Papua New Guinea, 1998**: Caused by an undersea landslide, this tsunami resulted in waves up to 15 meters high and killed more than 2,200 people.

9. **Sanriku, Japan, 1896**: Known for its incredibly high run-up heights, the tsunami resulted from an undersea earthquake and affected the Sanriku coast of Japan, killing over 22,000 people.

10. **Chile, 1960**: Triggered by the most powerful earthquake ever recorded (magnitude 9.5), this tsunami affected the entire Pacific, causing deaths as far away as Hawaii, Japan, and the Philippines.

Each of these historical tsunamis serves as a stark reminder of the immense power and potential devastation that this natural phenomenon can cause. Understanding these events can help improve preparedness and response strategies for future tsunamis.

9.1.6 Tsunami Glossary

Understanding tsunamis is easier when you know the terms that scientists use when discussing them. Here is a list of tsunami vocabulary terms and their definitions:

- Wave Train: A series of waves traveling together, separated by a relatively consistent distance, typically found in a tsunami event.
- Run-up: The maximum vertical height a tsunami wave reaches as it moves inland from the coastline.
- Tsunamigenic: Refers to any geological or cosmic event capable of producing a tsunami.
- Wavelength: The distance between two corresponding points on adjacent waves, such as from crest to crest or trough to trough.

- Wave Height: The vertical distance from the crest (top) of a wave to the trough (bottom).
- Wave Period: The time it takes for a single wave to pass a fixed point.
- Wave Frequency: The number of waves passing a fixed point per unit of time, often measured in Hertz (Hz).
- Wave Speed: The speed at which a wave travels, often calculated by multiplying the wave's frequency by its wavelength.
- Amplitude: The maximum displacement of the water surface from its rest position, essentially half of the wave height.
- Crest: The highest point of a wave.
- Trough: The lowest point of a wave.
- Drawback: The noticeable retreat of ocean water along the shore, exposing the sea floor, which often occurs just before a tsunami hits.
- Shoaling: The process where a wave's height increases as it enters shallower water.
- Refraction: The bending of a wave as it moves into areas with different depths, often causing the wave to align more parallel with the shoreline.
- Seismicity: Earthquake frequency, distribution, and magnitude within a specific region.
- Subduction Zone: An area where one tectonic plate is pushed below another, often the site of tsunamigenic events.
- Seismograph: An instrument that records the vibrations of the Earth, used to detect earthquakes and, by extension, potential tsunamis.
- Seismic Waves: The waves of energy caused by the sudden breaking of rock within the Earth or an explosion, which are the primary cause of earthquakes.

- Plate Tectonics: The scientific theory that describes the movement of Earth's lithosphere (crust and upper mantle) as divided into several large and small pieces known as tectonic plates.
- Aftershock: A smaller earthquake that occurs in the same general area during the days to years following a larger earthquake or "mainshock."
- Buoyancy: The ability of an object to float in water or another fluid, utilized in the design of tsunami-detecting buoys.

9.1.7 The science behind tsunamis

What is a tsunami?

A tsunami is a series of waves triggered by large-scale and sudden displacement of the ocean. Strong earthquakes at the bottom of the ocean or near the ocean are the most common causes, but landslides, volcanic activity, certain types of weather phenomena, and meteorites can also trigger tsunamis as shown in figure 5.

Tsunamis spread from their source in all directions, and are able to cross entire ocean basins, bypassing islands, and entering bays, straits, and rivers. When a tsunami arrives on the coast, it can trigger dangerous coastal flooding and strong currents, which can have an impact on occan operations and navigation that can last for hours or days.

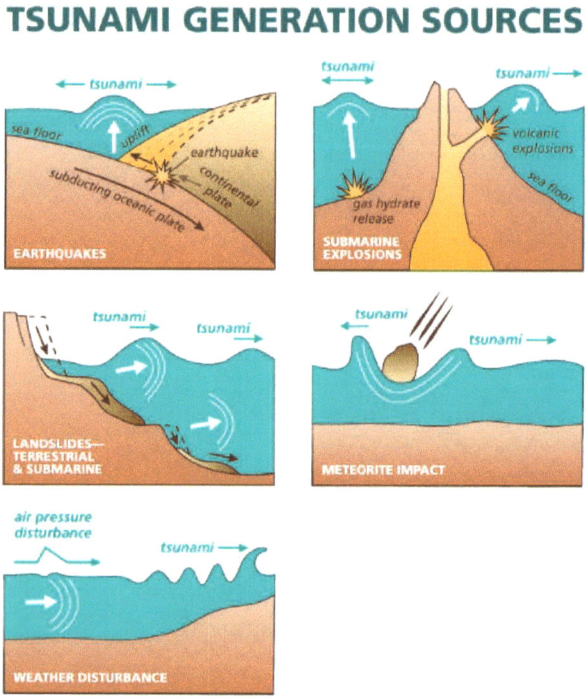

Figure 5 science behind tsunamis

Tsunamis are infrequent but can pose a serious threat to life and property when they occur. Tsunamis have claimed hundreds of thousands of lives and caused hundreds of billions of dollars in damage around the world.

In deep ocean water, tsunami waves may go unnoticed. But as the waves travel closer to the shore, they build in height as the water becomes shallower. The speed of tsunami waves depends on ocean depth: The deeper the water, the faster the wave travels. Tsunami waves may travel as fast as jet planes through deep waters, only slowing down when reaching shallow waters. While tsunamis are often referred to as tidal waves, this name is discouraged by oceanographers because tides have little to do with these destructive waves — though a tsunami at high tide may cause more damage and flooding.

9.1.8 What causes a tsunami?

About 80% of all known tsunamis are triggered by earthquakes. These seismic events move Earth's surface, displacing the water above and generating waves that rapidly travel in all directions across the ocean or body of water.

Not all earthquakes create tsunamis. An earthquake must be big enough and close enough to the ocean floor to cause the vertical movement of the ocean floor that typically sets a tsunami in motion. As the ocean floor rises or drops, so does the water above it. As the water moves up and down, seeking to regain its balance, the tsunami radiates in all directions. The amount of movement of the ocean floor, the size of the area over which it occurs (which may be reflected in how long the earthquake lasts) and the depth of the water at its source are all critical factors in the size of a resulting tsunami. Earthquakes can also cause landslides that generate tsunamis.

9.1.9 How landslides generate tsunamis

Tsunamis can be generated when a landslide enters the water and displaces it from above (subaerial) or when water is displaced ahead of and behind an underwater (submarine) landslide. Tsunami generation depends on the amount of landslide material that displaces the water, the speed it is moving and the depth it moves to. Landslide-generated tsunamis may be larger than seismic tsunamis near their source and can impact coastlines within minutes with little to no warning, but they usually lose energy quickly and rarely affect distant coastlines.

Most landslides that generate tsunamis are caused by earthquakes, but other forces (like gravity, wind and increased precipitation) can cause overly steep and otherwise unstable slopes to suddenly fail. Earthquakes that are not large enough to directly generate a tsunami may be large enough to cause a landslide, which can in turn generate a tsunami. A landslide-generated tsunami may occur independently or along with a tsunami directly generated by an earthquake, which can complicate the warning process and compound the losses.

9.1.10 What is a meteotsunami?

Large storms over the ocean or large water bodies can cause meteotsunamis, which, like tsunamis, can also be destructive. Meteotsunam is are driven by air-pressure disturbances often associated with fast-moving weather events, such as severe thunderstorms, squalls, and other storm fronts. The storm generates a wave that moves towards the shore, and is amplified by a shallow continental shelf and inlet, bay, or another coastal feature. Meteotsunamis have been observed to reach heights of 6 feet (1.82 meters) or more. They occur in many places around the world, including the Great Lakes, Gulf of Mexico, Atlantic Coast and the Mediterranean and Adriatic Seas. Identifying a meteotsunami is a challenge because its characteristics are almost indistinguishable from a seismic tsunami. Meteotsunamis can also be confused with wind-driven storm surge or a seiche. These uncertainties make it difficult to predict a meteotsunami and warn the public of a potential event. However, NOAA scientists have identified atmospheric conditions that are likely to generate meteotsunamis and continue to research them.

9.1.11 Where do tsunamis occur?

Although tsunamis occur relatively infrequently, they can be a serious threat to life and property. Tsunamis have claimed hundreds of thousands of lives and caused hundreds of billions of dollars in damage around the world.

A tsunami can strike any coast (ocean, lake or even river) at any time. There is no season for tsunamis. This is quite evident when we look at records of past tsunamis. Tsunamis typically happen where there are large faults that can generate large earthquakes. Most of these large faults are around the Pacific Ring of Fire. However, any active fault that crosses water may be capable of generating a tsunami. As mentioned above, other causes of tsunamis include landslides and volcanic eruptions which can happen far away from any ocean coastline.

Tsunamis are sometimes referred to as "tidal waves," but this is very misleading as they are not related to tides.

9.1.12 Tsunamis throughout history

What information do we have about tsunamis that happened in the past?

The NOAA/World Data Service (WDS) tsunami database is a listing of historical tsunami source events and runup (inland flooding extent) locations throughout the world that range in date from 2000 BCE to the present. The natural hazards datasets for tsunamis are available through the HazEL (Hazardous Event Lookup) interface, developed by NOAA's National Centers for Environmental Information (NCEI).

The events were gathered from scientific and scholarly sources, regional and worldwide catalogs, tide gauge data, deep ocean sensor data, individual event reports and unpublished works. One example, published on the USGS website, demonstrates how old records in Japan helped unravel what happened during the "orphan" tsunami of 1700.

View of Krakatoa during the earlier stage of the eruption, from a photograph taken on Sunday, 27 May 1883. (Image credit: Published in the Report of the Krakatoa Committee of the Royal Society.)

Figure6 Japan- during the "orphan" tsunami of 1700.

9.1.13 2004 Indian Ocean tsunami

On December 26, 2004, the third-largest earthquake recorded in the world (since 1900) occurred off the west coast of the Indonesian island of Sumatra. One of the most devastating and catastrophic tsunamis recorded, it spread outward throughout the Indian Ocean and struck the coastlines of 17 countries in southeastern and southern Asia and eastern and southern Africa.

In total, just over a quarter of a million people lost their lives from this tragic event, with another 1.7 million people displaced. In 2017, the cost and damage from the tsunami was estimated to be approximately $13 billion.

Figure 7 Indian Ocean tsunami

Shown in figure Banda Aceh, Sumatra, Indonesia - January 4, 2005:

A lone mosque remains standing among the damage of a coastal village near Banda Aceh, Sumatra, Indonesia following the 2004 Indian Ocean tsunami. (Image credit: Getty images)

The 9.1 magnitude earthquake in the Indian Ocean was about 19 miles below the ocean floor, and the length of the rupture was roughly 800 miles — basically the size of California's coastline. The quake displaced a massive volume of water to generate a global tsunami, which reached some areas within 20 minutes and others in 7 hours. The tsunami was observed by more than 100 coastal water-level stations in the Atlantic and Pacific ocean basins. One of the hard-to-grasp statistics from the tsunami were wave heights that reached 167 feet in Indonesia's Aceh province in northern Sumatra, which resulted in flooding up to 3 miles inland.

Once the tsunami reached portions of Asian and African coastlines, the waves advanced inland, carried debris, destroyed once-thriving communities and ruined many livelihoods. Imagine: No more homes, basic services, and critical infrastructure such as roads, power lines, bridges, clean water, gasoline/fuel — essentially all the things we rely on in our modern world. The economies of many countries were devastated. The saltwater from the tsunami damaged soils, crops and vegetation, while completely changing the overall environmental landscape of many Indian Ocean coastal communities.

An order of magnitude hard to imagine

Since 1900, about 5% of all tsunamis have occurred in Southeast Asia, but the scope and magnitude of the 2004 tsunami set it apart from others.

Since this catastrophic event, NOAA has made significant improvements to its tsunami detection, forecasting, warning and preparedness capabilities, including:

- U.S. Deep-ocean Assessment and Reporting of Tsunamis (DART) buoys increased from 6 to 39.
- Installed or upgraded 188 coastal water-level stations to support tsunami warning operations.
- Upgraded the Global Seismographic Network to transmit 100% of its seismic data in real time.
- Developed forecasting models to predict the arrival time, duration, height and extent of flooding.
- Expanded the Tsunami Warning Centers' responsibilities.
- Started the TsunamiReady program to help communities minimize risk through better hazard assessment, planning, education and warning communications.

- In 2007, the U.S. Congress Established a new Act of Law, the Tsunami Warning and Education Act, which was later revised and reauthorized in 2017 as the Tsunami Warning, Education, and Research Act (TWERA), which outlines the responsibilities and goals for NOAA and a successful U.S. Tsunami Warning System and Program. TWERA formally established the National Tsunami Hazard Mitigation Program, a program led by NOAA to advance tsunami preparedness at the local level through partnerships and grant funding for at-risk states and territories.

9.1.14 Monitoring and alert systems

Where do tsunamis strike?

Tsunamis can strike any U.S. coast and territory any time. The coastlines most at risk are in the Pacific Ocean basin and the Caribbean Sea shown in figure 8.

NOAA has two tsunami warning centers in the U.S. that are staffed 24 hours a day, 7 days a week, to monitor and alert for tsunamis from earthquakes. Scientists detect the earthquake, observe the tsunami, forecast tsunami impacts, issue tsunami alert messages, conduct public outreach and coordinate with partners to continually improve warning operations with the main mission in protecting life and property from tsunamis.

Figure 8 Monitoring and alert systems

Tsunami buoy systems quickly provide critical information to warning centers. (Image credit: NOAA)

The National Tsunami Warning Center, located in Palmer, Alaska, serves the continental U.S. (East, West and Gulf Coasts), Alaska and Canada. In Ford Island, Hawaii, the Pacific Tsunami Warning Center (PTWC) serves the Hawaiian Islands and the U.S. Pacific and Caribbean territories, including American Samoa, Guam, Northern Mariana Islands, Puerto Rico, U.S. Virgin Islands and the British Virgin Islands. PTWC also serves as the Pacific Tsunami Warning System's main tsunami service provider. In this role, PTWC works closely with other international, subregional and national centers to monitor seismic and sea level stations around the Pacific Ocean for large earthquakes and tsunami waves. Both tsunami warning centers are part of NOAA's National Weather Service (NWS). NWS also leads the National Tsunami Hazard Mitigation Program (NTHMP) and the International Tsunami Information Centeroffsite link (ITIC). The NTHMP is a coordinated U.S. national effort to mitigate the impact of tsunamis through public education, community response planning, hazard assessment and warning coordination.

How does NOAA monitor and detect tsunamis generated by earthquakes or volcanic eruptions?

NOAA has a suite of coastal water-level stations along the West, East and Gulf Coasts and surrounding Alaska, Hawaii, Puerto Rico and U.S. Territories. These stations collect important data about the height of the ocean at specific coastal locations and monitor tides. The data is then relayed by satellite to NOAA's warning centers, where it is used to confirm tsunami height and arrival time, and is incorporated into tsunami forecast models.

In addition to vast coastal water-level stations, NOAA has 39 Deep-ocean Assessment and Reporting of Tsunami (DART®) sea-level systems located strategically throughout the Pacific and Atlantic Ocean basins, the Gulf of Mexico and the Caribbean Sea. The DART systems help identify early detection, measurement and real-time reporting of tsunamis in the open ocean waters. Data from these coastal and DART stations are then relayed via satellite to the warning centers, where they are used to confirm tsunami height and arrival time and incorporated into tsunami forecast models. A constellation of satellite radar altimeters — including NOAA's partnered missions (Jason-3 and Sentinel-6) — measure sea surface height and roughness. Those measurements provide subtle but detectable signals as tsunamis cross the deep ocean: A small, rapid change in sea level paired with altered surface roughness. The data can support and refine tsunami propagation models, allowing for more accurate predictions of wave travel times and coastal impacts. Additionally, altimeter data can enhance bathymetric models by providing information on ocean bottom topography, which affects tsunami wave speed and direction. This approach is not yet operational at NOAA's Tsunami Warning Centers.

When NTWC and PTWC issue tsunami alerts, what do they mean?

The National Tsunami Warning Center (NTWC) and Pacific Tsunami Warning Center (PTWC) issue four different messages and alerts to notify the public, emergency managers and other local officials and partners about the potential for a tsunami following a possible tsunami-generating event show in figyre 9.

Below are the four types of messages and their definitions:

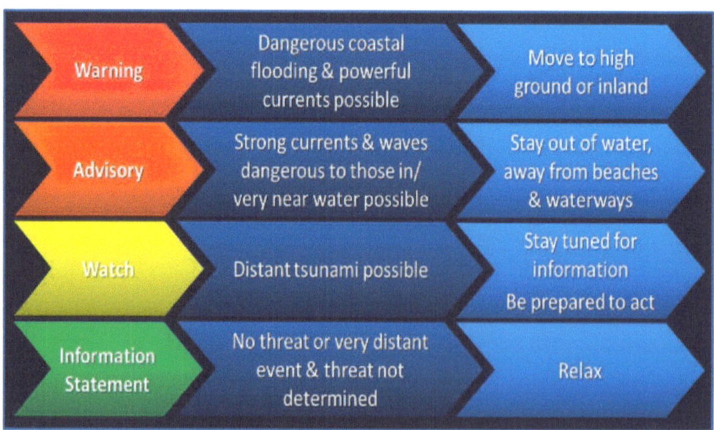

Figure 9 tsunami alerts process

Tsunami alerts for U.S. and Canada: what they mean and what you should do (full tsunami alert definitions).

Download Image

9.1.15 Tsunami mitigation and research: What is next?

This action aims to provide consistent tsunami forecasting, alerting, decision-support services and seamless, comprehensive service backup. Substantial progress has been made in recent years, solidifying the requirements for new common analytical software to common operating systems and technical support. The most critical component of the effort to establish a common operating platform is to provide service backup between the two Tsunami Warning Centers if one Tsunami Warning Center becomes inoperable or requests back-up.

Figure 10 Tsunami mitigation

Satellite imagery of the Tonga-Hunga Ha'apai volcano in the South Pacific Ocean on January 15, 2022. Shown in figure 10NOAA's Tribal Relations Team works to strengthen and build relationships with tribes, Alaska Native communities and Native Hawaiian and Pacific Islanders, and works to ensure accountability (by integrating tribal treaties and reserved rights into agency decision-making processes, and upholding Executive Orders), maintain and establish relations and promote awareness of the policies, activities and concerns related to these vital partners. Congress has authorized NOAA to lead the National Tsunami Hazard Mitigation Program (NTHMP) through

NWS's Tsunami Program. Tasked with reducing the impact of tsunamis on the nation, the NTHMP is a unique and effective partnership between the NOAA, the Federal Emergency Management Agency (FEMA), the U.S. Geological Survey (USGS) and 28 U.S. coastal states and territories. One of the greatest strengths of the NTHMP is the extraordinary commitment each member brings to the program and the high level of cooperation between federal, state and territory emergency management and scientific partners. This collaboration enhances cost-effectiveness and improves the program's ability to implement consistent national policies and projects at the local level. The NTHMP has four strategic overarching priorities or themes:

1. Hazard and Risk Assessment
2. Education and Preparedness
3. Mitigation and Recovery
4. Alert, Warning, and Response

The themes are supported by goals and strategies, which the NTHMP will strive to meet through collaboration between partners and stakeholders.

New frontiers in tsunami research

The scientists at NOAA's Pacific Marine Environmental Laboratory (PMEL) in Seattle are working with NWS staff to develop new tsunami detection and modeling methods. The NOAA Center for Tsunami Research at PMEL aims to translate new science into a more accurate, fast and actionable tsunami forecast capability for NOAA's Tsunami Warning Centers.

PMEL tsunami research mainly focuses on providing accurate inundation forecasts in real-time before flooding from a tsunami reaches threatened coastlines. If timely and accurate, the tsunami inundation estimates are critical for saving coastal residents' lives during a tsunami event. Modeling, real-time tsunami detection and data assimilation are central to PMEL's tsunami research.

About 80% percent of all known tsunamis are triggered by earthquakes. That is why the current tsunami warning system is focused on earthquake-generated tsunamis. However, landslides, volcanic eruptions and asteroid impacts have caused catastrophic tsunamis. Large storms over the ocean or large water bodies can cause meteotsunamis, which can also be destructive. Specialized models developed by NOAA's National Center for Tsunami Research can predict coastal inundation from these other tsunamis, too, provided they get accurate information about the size and location of the triggering event.

Data from atmospheric and weather observations combined with real-time coastal and deep-water tsunami detection, as well as AI-assisted analytics, could provide enough information for models to forecast wave heights before a non-seismic tsunami arrives onshore.

Additionally, NOAA is working with emergency managers to assess long-term tsunami hazards by modeling hypothetical events and to prepare tsunami inundation maps under different scenarios to build tsunami-resilient communities.

NOAA is collaborating with the American Society of Civil Engineersoffsite link to develop the world's first probability-based tsunami design provisions. Among their goals:

- Establishing building codes requiring structures in inundation zones to be able to withstand wave forces.
- Assessing tsunami debris hazards. One of the most destructive elements of a tsunami results from the debris that is swept up and propelled shoreward, then seaward, by powerful waves.
- Creating new tsunami inundation (flooding) maps that are based on probabilities of occurrence for different size earthquakes which can be tailored to a given location and take into account hazards amplified by sea level rise and changing coastlines.

10.1 Classification Natural Hazards| Class 11

Understanding the different classifications and types of natural hazards and disasters is essential for studying geography. These phenomena cover a broad spectrum of occurrences, including atmospheric disturbances like hurricanes and floods as well as geological upheavals like earthquakes and volcanic eruptions. Examining these divisions sheds light on the dynamic interactions that shape our world and helps develop mitigation and preparedness plans for disasters.

What is a Natural Hazard?

Conditions found in the natural world that have the potential to endanger persons, property, or both are known as natural hazards. These could be sudden or enduring elements of the corresponding environmental settings, such as ocean currents, the Himalayas' steep slopes, fragile structural characteristics, or the harsh weather in deserts or glaciated regions.

Classifications of Natural Disaster

There are four different types of classifications of natural disaster

1. Atmospheric Classification
2. Terrestrial Classification
3. Aquatic Classification
4. Biological Classification

10.1.1 Atmospheric Classification

Earth's atmosphere is the primary location for natural disasters, which are the focus of atmospheric classification. These catastrophes are caused by conditions and atmospheric circumstances.

Examples of Atmospheric Classification

Atmospheric Disasters	**Examples**
Storms	Thunderstorms, Hailstorms, Blizzards
Tropical Cyclones	Hurricanes, Typhoons, Cyclones
Tornadoes	Violent tornadoes, Twisters
Droughts	Prolonged dry spells, Water scarcity
Extreme Temperature	Events Heat Waves, Cold Waves, Frost Events

10.1.2. Terrestrial Classification

Natural catastrophes that affect land surfaces, such as geological and geomorphological hazards, are the main emphasis of the terrestrial classification.

Examples of Terrestrial Classification

Terrestrial Disasters	Examples
Earthquakes	Seismic Events, Tremors
Volcanic Eruptions	Lava Flows, Ashfall, Pyroclastic Flows
Landslides	Rockfalls, Mudslides, Debris Flows
Avalanches	Snow Avalanches, Ice Avalanches
Soil Erosion	Wind Erosion, Water Erosion

10.1.3 Aquatic Classification

Natural catastrophes that happen in bodies of water, such as lakes, rivers, seas, oceans, and groundwater, are the focus of the aquatic classification.

Examples of Aquatic Classification

Aquatic Disasters	Examples
Floods	Flash Floods, River Flooding, Coastal Flooding
Tidal Waves (Tsunamis)	Seismic Tsunamis, Landslide Tsunamis
Ocean Currents	Gulf Stream, Kuroshio Current
Storm	Surge Coastal Storm Surges, Seawater Inundation

10.1.4. Biological Classification

Natural catastrophes that affect living things and ecological systems are categorized as biological classification.

Examples of Biological Classification

Biological Disasters	Examples
Plants and Animals as Colonisers	Ecological Succession, Pioneer Species
Insects Infestation	Locust Swarms, Bark Beetle Outbreaks
Fungal, Bacterial, and Viral Diseases	Bird Flu Outbreaks, Dengue Fever Epidemics

11.1 Floods in India, Causes, Impacts, Mitigation, Measures

Floods are natural disasters that occur when a large amount of water submerges normally dry ground. India is highly vulnerable to floods in which Assam, Bihar, Odisha, Uttar Pradesh, and West Bengal are largely affected.

Floods are the most common form of natural disaster, occurring when a large amount of water submerges normally dry ground. Floods are frequently produced by excessive rainfall, quick snowmelt, or storm surge from a tropical cyclone or tsunami in coastal locations. India is highly vulnerable to floods in which Assam, Bihar, Odisha, Uttar Pradesh, and West Bengal are largely affected. Flood disasters affect people and the impact of floods cannot be completely prevented, while measures can be taken to mitigate the impact and ensure that people are prepared to deal with the consequences of the flood hazards.

11.2 Flood-prone Areas of India

India is the worst-affected country by floods after Bangladesh, accounting for one-fifth of all flood-related deaths worldwide. According to the Ministry of Jal Shakti, the recent estimate on flood-prone areas in the country is 49.15 Mha. The annual average cropped area affected by flood is approximately 3.7 million hectares.

12.1 Indo-Gangetic- Brahmaputra plains:

The most flood-prone areas in India are the Brahmaputra and the Ganga River basins in the Indo-Gangetic- Brahmaputra plains in North and Northeast India, which carry 60 percent of the nation's total river flow.

Affected states: Nearly 75% of total Indian rainfall occurs during a brief monsoon season (June-September). As a result, the rivers witness a heavy discharge during these months, leading to widespread floods in Uttar Pradesh, Bihar, West Bengal, and Assam.

Assam Valley is highly flood-prone, primarily due to heavy rains and silting of tributaries like Dihang and Lohit.

In Kashmir Valley, the Jhelum is unable to carry the river discharge.In Punjab and Haryana, the problem is primarily due to inadequate drainage.The Central Indian and peninsular river basins, covering Madhya Pradesh, Orissa, Andhra Pradesh, Maharashtra, and Gujarat, contain the Tapi, Narmada, and Chambal rivers, which often experience excess rainfall.In Andhra Pradesh, the Kolleru lake submerges vast areas along its fringes and cyclonic storms bring heavy flooding in the coastal areas.

12.1.2 Flash flood-prone areas:

In recent decades, states such as Rajasthan, Gujarat, Haryana, and Punjab have been inundated by flash floods.

12.1. 3 Causes of Floods in India

India is prone to floods due to its geography, heavy rainfall during the monsoon and global warming leading to more frequent and intense rainfall events, etc. Floods are caused by both natural as well and anthropogenic factors.

12.1.4 Natural Causes

Heavy rainfall:

Heavy rain in the catchment area of a river causes water to overflow its banks, which results in the flooding of nearby areas.During the monsoon season, there are many events of flood in northern and northeastern India and the Gangetic plains due to heavy rainfalls.

Cloud Burst:

Cloud bursts lead to sudden and intense rainfall, often resulting in flash floods, particularly in the Himalayan regions. This phenomenon contributes to significant sediment deposition in riverbeds, which diminishes the rivers' capacity to carry water. Over time, natural processes like erosion and meandering can alter river channels, further increasing the risk of flooding. The Kosi River in Bihar exemplifies this issue, earning the nickname "sorrow of Bihar" due to its frequent course changes and associated floods. Additionally, India's coastal areas, especially those bordering the Bay of Bengal and the Arabian Sea, are vulnerable to tropical cyclones that can trigger severe flooding through heavy rains and storm surges. A recent example is Cyclone Michaung, which struck Andhra Pradesh in December 2023, resulting in devastating storm-induced floods that caused loss of life and extensive property damage.

Himalayan glacier melt:

The melting ice can result in the formation of glacial lakes, which pose a risk of glacial lake outburst floods (GLOFs). Example: Uttarakhand Glacial burst.

Climate Change:

Climate change has aggravated the situation of flood risks. In mountainous regions, three effects of climate change are there -more intense precipitation due to a warmer atmosphere, shifting snow and rain patterns, and the effects of wildfires on the landscape Climate change is also causing intense rain and the occurrence of tropical rainfall in the Indian Ocean region. Thus, it has also caused an increased risk of floods in the Western Ghat region. Anthropogenic Causes

Deforestation:

Vegetation plays a crucial role in absorbing rainwater and reducing surface runoff. Deforestation disrupts natural rainwater absorption and reduces surface runoff, increasing runoff volumes and river flooding risk.According to a study, with every 10 percent increase in trees being cut down, the flood risk increases by up to 28 percent.

Poor drainage: Urban growth converts natural land into impermeable surfaces. This reduces land's ability to absorb water and leads to increasing surface runoff, and insufficient or poorly maintained drainage systems increase flood risk.For instance, recent floods in Bangalore are more attributed to the reasons for poor drainage and urbanisation.Dam and embankment failure: Dam collapse and river embankment failures due to poor infrastructure can also cause floods.In case of heavy rainfall and limitation of water storage capacity of dams, the dam gates are opened which causes severe floods in the nearby areas.

Encroachment on floodplains:

Construction and settlement in floodplain areas disrupt the natural flow of water during floods.Encroachment reduces the capacity of floodplains to store excess water, causing it to overflow into surrounding areas.

Encroachment of water bodies:

Encroachment of lakes, ponds and other water bodies aggravates the risk of floods because they cannot contain the extra water when the situation of flooding arrives.

12.1.5 Impacts of Floods in India

Floods arc generally destructive to humans and the natural environment, but they also contribute positively in some cases.

Negative Impact

Immediate impacts: It causes loss of life and property, crop destruction, livestock loss, infrastructure malfunction, disruption in movement and deterioration of health conditions due to waterborne diseases.According to Central Water Commission (CWC) data, in the last 65 years, floods killed 109,412 people, and 258 million hectares of crops were damaged.

Economic loss: Floods cause damage to infrastructure, leading to livelihood loss and disruption in industries and businesses, causing production delays, supply chain disruptions, and financial losses.According to SBI research, floods in North India caused an estimated economic loss worth ₹10000-15000 crore in 2023.

Displacement: Floods result in a humanitarian crisis and displacement of the people. Flooding in Jammu and Kashmir in 2015, Uttarakhand in 2013, and Assam in 2012 displaced millions of people.Health risks and waterborne diseases: Floods contaminate water sources, leading to an increased risk of waterborne diseases such as cholera, typhoid, and diarrhoea. The World Health Organization estimates that floods in India result in the loss of millions of productive days and contribute to a substantial disease burden.

Environmental impacts: Floods cause soil erosion, sedimentation, and degradation of ecosystems.Sediment-laden floodwaters can damage agricultural land, affect soil fertility, and impact biodiversity.

Positive Impact: Recharging water sources: Inundation of the flood plains helps recharge the groundwater, which is an important source of drinking water and is essential for agriculture.

Agriculture: Floodwaters carry nutrients and sediments, which are deposited on flood plains, enriching the soil. Rice paddies are flooded deliberately to take advantage of this natural fertilisation process.

Rejuvenation of the river ecosystem: The river ecosystem is crucial for the biodiversity of fish, wildlife, and waterfowl, and its seasonal variability and varied sediment and flow regimes help maintain this balance.

Fishery: Floods trigger aquatic species spawning and migration, providing abundant fish supply and alternative income sources at the household level.

12.1.6 India's Preparedness for Floods

The Indian government has implemented many initiatives and policies to lessen the impact of floods through various state and central disaster management support organisations.

Central Water Commission (CWC): It was set up in 1945 and focuses on flood control, water resource conservation, irrigation, hydropower generation, flood management, and river conservation in the country..

National Disaster Management Authority (NDMA): NDMA has prepared these Guidelines for Flood Management, to assist the ministries and departments of the GOI, the state governments, and other agencies in preparing Flood Management plans (FMPs).

The National Water Policy (2012): It emphasises flood control through structural measures, integrated reservoir operation, natural drainage system rehabilitation, farming systems, and non-agricultural development, aiming for long-term solutions to devastating floods.

Flood Forecasting Network: The CWC has implemented a flood forecasting system with 175 stations across major interstate rivers, enabling real-time data collection, automatic transmission, flood forecast formulation, and efficient information dissemination.

Flood Management and Border Areas Programme (FMBAP): It is being implemented throughout the country for effective flood management, erosion control, and anti-sea erosion and to help maintain peace along the border.

Structural Measures: Since 1954, the government has constructed 33,928 km of new embankments and 38,809 km of drainage channels, completed 2,450 town protection works, and raised 4,721 villages above flood levels.

Post-disaster response and recovery: India has set mechanisms for the post-disaster response, such as search and rescue operations, relief and rehabilitation camps, NDRF forces, and the use of earth observation satellites. Cooperation with the neighbouring countries: India has cooperation with neighbours like Nepal, China, and Bhutan for the exchange of hydro-meteorological data for early preparedness of flood management.

12.1.7 Challenges with Flood Management in India

Despite significant investments and ongoing flood management measures in India, socioeconomic damages and mortality tolls remain high. The following are the challenges of flood management in India:

Complex hilly regions: Flood risk management in hilly regions is still in the infancy stage, particularly due to complex and tough terrain with limited accessibility and a low level of monitoring.Technological problems: Innovative technologies are emerging to help manage flood risk, but these are not always straightforward to implement and technology alone will not address all our challenges.

Lack of policy implementation: Though non-structural measures (such as regulations, land use planning, flood zoning, and flood forecasting) find their mentions in flood-related policies and are practised in a few regions of the country, they are yet to be enacted extensively by the States.

Addition of new flood-prone areas: Flood-prone regions are growing rapidly due to climate and human factors, posing challenges for organizations to implement effective plans and policies to minimize human and economic losses during and after disasters.Policy failure: Over the years, the National Flood Commission recommendations have failed to achieve much success in mitigating floods, and the data show that flood damage and fatalities have only increased for years.Suggested Measures.A NITI Aayog report on flood management under Rajiv Kumar has suggested effective and long-lasting strategies for flood control and management that involve structural and non-structural measures along with the use of modern technologies.

Non-structural measures:

Implementation of floodplain zoning: It has been suggested to implement the floodplain zoning approach as mandated by the Ministry of Jal Shakti.

Scientific research: It has called for modernisation in the collection of hydrometeorological data, flood forecast formulation, forecast dissemination and focus on scientific research in the development of a Model-based system to forecast flash floods.

Use of space technology: Emphasis is to be laid on the application of space technology and the use of state-of-the-art technology in spatial flood early warning and near real-time monitoring and mapping of floods.

Structural measures:

Storage reservoirs: The construction of large storage reservoirs that moderate flood peaks is the need of the hour.

Embankments: The raising and strengthening of existing embankments and creating new embankments, channelisation, drainage & channel improvement, and anti-erosion works have been suggested.

Interlinking of rivers: The projects for interlinking of rivers for diversion of flood water to water-scarce regions should be completed.

13.0 Question related to disaster management.

Disaster Management MCQs

1. In which Phase of Disaster damage Assessment is done?

   A) Warning phase

   B) Impact Phase

   C) Rescue phase

   D) Re-Habilitation Phase

Answer: D

2. In relation to Disaster Preparedness, which one of the following is more serious from response point of view?

   A) 1st alert

   B) 2nd alert

Answer: B

3. A physical situation which may cause human injury, damage to property or the environment, is called

   A) Hazard

   B) Risk

Answer: A

4. The likelihood of an undesired event occurring within a particular period or under specified circumstances is known as

   A) Hazard

   B) Risk

Answer: B

5. Which of the following is the main contributory cause for Pipeline Disasters in the world?

   A) Third-party Activity

   B) Corrosion

   C) Mechanical failure

   D) Operational failures

Answer: A

6. Root Cause Analysis of Disasters can be done through which of the following methods?

   A) HAZOP

   B) Checklist

   C) Consequence Analysis

   D) Why-Why

Answer: D

7. 'Blow-Out' is a Natural Disaster.

    A) Yes

    B) No

Answer: B

8. The principle which takes into account the precautions to be taken being on safer side and avoidance of taking undue risks from hazards, the damage potential of which is not yet fully known, is known as:

    A) Guiding Principle

    B) Risk Assessment Principle

    C) Hazard Analysis Principle

    D) Precautionary Principle

Answer: D

9. The agreement or understanding that has been reached among various nearby industries to help each other during emergency is called:

    A) Memorandum of Agreement

    B) Agreement of Mock-exercise

    C) Industrial Charter

    D) Mutual-Aid Agreement

Answer: D

10. How many Safety Integrity Levels exists?

    A) 1

    B) 2

    C) 3

    D) 4

Answer: D

11. Which of the following SIL levels represents Catastrophic impacts of a Disaster?

    A) SIL-1

    B) SIL-2

    C) SIL-3

    D) SIL-4

Answer: D

12. What could be the Primary and Secondary Protections for protection of a vessel against Over-pressure situation?

    A) PSH and PSV

    B) PSL and ESS

    C) LSL and ESD

    D) FSL and FSV

Answer: A

13. In a Oil and Gas Drilling rig, an area within 4.5 m of the shale shaker in open air is considered as:

    A) Zone-1

    B) Zone-2

    C) Zone-0

Answer: A

14. The area within 3 meter from the discharge nozzle of the relief valve is considered as:

    A) Zone-0

    B) Zone-2

    C) Zone-1

Answer: B

15. Which of the following will be found on a Material Safety Data Sheet?

    A) Spill and leak procedures

    B) Health Hazard Information

    C) Special protection Information

    D) All of the above

Answer: D

16. Section 2 of an MSDS will have information about

    A) Hazardous ingredirents

    B) Fire and Explosion data

    C) Reactivity data

    D) Manufacturers' data

Answer: A

17. In case of an emergency involving a chemical or product, an emergency telephone number can be found in which section of the MSDS?

    A) Section 1

    B) Section 2

    C) Section 5

    D) Section 8

Answer: A

18. The yellow portion of a NFPA diamond represents:

    A) Health Hazard

    B) Flammability Hazard

    C) Reactivity Hazard

    D) Special Hazard Information

Answer: B

19. The ill-health effect caused by one-time, sudden, high exposures is often called:

    A) Chronic effect

    B) Acute effect

    C) Systematic effect

Answer: B

20. Monitoring the levels of exposure to toxic substances is medically done through analysis of body fluids e.g. blood, urine, expired air etc. The process is known as:

    A) Condition monitoring

    B) Biological conditioning

    C) Exposure monitoring

    D) Biological monitoring

Answer: D

21. The ratio of Amount of toxic chemical to the IDLH value of the chemical is called:

    A) Fire Load

    B) Toxic Load

    C) Pollution Load

    D) Toxicological Load

Answer: B

22. Which of the following steps of Toxic Risk Assessment exercise considers the possible routes of entry of chemicals into body and its reaction thereby?

    A) 1st step

    B) 2nd step

C) 3rd step

D) 4th step

Answer: C

23. About 2 ppm of Chlorine concentration in Air causes?

A) Irritation of eyes and nose

B) Irritation of throat

C) Intense cough & fits

D) Immediate death

Answer: A

24. If 'Cash' is a hazard, theft is the consequence, then if 'electricity' is the hazard, what could be the consequence ?

A) Damage

B) Hearing loss

C) loss of water

D) shock

Answer: D

25. What is the purpose of hazard identification in Disaster preparedness?

A) to identify the probable causes of disasters

B) to identify the probable disaster scenarios

C) to quatify the consequences

D) to suggest mitigation measures

Answer: B

26. 'Pareto Analysis' is a:

A) Risk quantification technique

B) Risk analysis technique

C) Risk mitigation technique

D) Risk prioritization technique

Answer: D

27. 'HAZID' is done to carry out safety hazard identification, what is it known as which is used for identification of environmental issues?

A) ENVID

B) WBCSD

C) ENVDI

D) Sustainability

Answer: A

28. Atmospheric Stability Class D corresponds to which of the following weather conditions?

A) Slightly Stable

B) Stable

C) Highly unstable

D) Neutral

Answer: D

29. An estimated Risk of 2×10^{-4} per year means:

A) One chance in 100 years

B) One chance in 10000 years

C) Two chances in 10000 years

D) Two chances in every 1000 years

Answer: C

30. When unacceptable risk is brought under the acceptability criteria through additional risk reduction measures, the identified risk is said to be within

    A) As low as Regularly Possible (ALARP) zone

    B) As low as Reasonably Practicable (ALARP) zone

    C) As low as Reasonably Possible (ALARP) zone

    D) As low as Rationally Probable (ALARP) zone

Answer: B

31. In Risk Analysis, Vulnerability Model represents:

    A) How many persons will be effected due to exposure

    B) How much area will be effected due to the event

    C) How long the effect will last in the community

    D) How probable that the community will be effected

Answer: A

32. Under damage criteria due to explosion pressure wave, 0.3 bar over pressure represents that:

    A) 10% houses will get damaged

    B) 90% houses will be seriously damaged

    C) Windows of houses will be smashed

    D) Glass of doors and windows will break

Answer: B

33. The catastrophic failure of a pressure vessel due to intense heating from a torch fire or a severe mechanical impact is commonly known as:

    A) UCVCE

    B) CE

C) BLEVE

D) Cascading effect

Answer: C

34. In Risk Assessment, 20/80 principle means:

A) 80% of total risk comes from 20% of cases

B) 20% of total risk originates from 80% of cases

Answer: A

35. Which one of the following has more damaging effect after a liquefied flammable gas release?

A) Early Ignition

B) Delayed Ignition

Answer: B

36. In Layer of Protection (LOP) Analysis, which of protection covers an 'onsite Emergency Plan'?

A) 1st Layer

B) 3rd Layer

C) 5th Layer

D) 4th Layer

Answer: D

37. Which of the following is a hazard quantification method?

A) FMEA

B) Fault-tree

C) HAZOP

D) What-if Analysis

Answer: B

38. Cause-Consequence Analysis (CCA) is a combination of:

    A) Fault-tree & Event tree Analyses

    B) Fault-tree & HAZOP Analyses

    C) MCA & Event tree Analyses

    D) MCA & HAZOP Analyses

Answer: A

39. The roles and responsibilities identified for disaster responses under Disaster Management Plan is known as:

    A) Incident Organogram

    B) Disaster Response Structure

    C) Incident Command System

Answer: C

40. The responsibility of press briefing during a disaster event normally rests on:

    A) Incident Controller

    B) Public relation Controller

    C) Transport Controller

    D) Security Controller

Answer: B

41. Whose responsibility normally it is to declare an Onsite Emergency in an industry?

    A) Incident Controller

    B) Site Main Controller

    C) Key person (Fire and safety)

Answer: B

42. Under which section of the Factories Act, 1948 (amended to 1987), an occupier of a factory has to formulate an Onsite Emergency Plan?

    A) 41-B(4)

    B) Section 13 – B(2)

    C) Section 46 – B(5)

Answer: A

43. Indian Coast Guard has capabilities for responding to which tier of Oil spill contingencies?

    A) Tier-I

    B) Tier-II

    C) Tier-III

Answer: B

44. Which of the following techniques are not used for Oil spill response?

    A) Use of booms and Skimming

    B) Use of dispersants

    C) Burning

    D) Bio-recovery

Answer: B

45. A hazardous chemical can be identified by which of the following numbers?

A) ID Number

B) NFPA Number

C) UN Number

D) Hazard index Number

Answer: C

46. Under various zones created around an incident site for effective response, which one normally used for command, control and support mobilization purposes?

A) Cold zone

B) Hot zones

C) Warm zone

Answer:

47. Within which zone "Emergency Control Center" can be situated?

A) Hot zone

B) Support zone

C) Warm zone

Answer: B

48. 'THERP' study is related to:

A) Risk Perception

B) Safety Analysis

C) Human error

D) Work study

Answer: C

49. Which of the following is not a technological factor used for Vulnerability assessment?

A) Process Safety

B) Individual difference

C) Communication failure

Answer: B

50. Which of the following is a human error in Disaster Management model?

A) Error of Omission

B) Typographical errors

C) Error of Commision

D) All of the above

Answer: D

51. Secondary fires are caused due to which of the following phenomenon?

A) Flash back

B) Burn-back

C) Fire spread

D) domino effect

Answer: B

52. What is the safe level of heat radiation (in Kw/m^2)for an average human being?

A) 1.75

B) 2.8

C) 6.4

D) 14.4

Answer: A

53. A Deep burn, characterized by destruction of all skin layers is termed as:

A) 1st degree burn

B) 2nd degree burn

C) 3rd degree burn

Answer: C

54. A Probit equation used for calculation of effects due to:

    A) Flammability

    B) Toxicity

    C) both Flammability and toxicity

Answer: C

55. In Risk estimation, a F-N curve is referred to depict which of the following risks?

    A) Individual Risk

    B) Group Risk

    C) Process Risk

Answer: B

56. The uncontrolled sudden violent escape of well fluids during oil and gas exploitation activities is known as:

    A) Blow out

    B) Burn out

    C) explosion

    D) Blast out

Answer: A

57. Which of the following classifications of blowout causes minor level of pollution?

    A) Class I

    B) Class II

    C) Class IV

    D) Class V

Answer: D

58. A normally directional well drilled to establish a direct communication with the well under blow out, is known as:

    A) New well

    B) Rehabilitatory well

    C) Relief well

    D) killing well

Answer: C

59. Lethality (Pr) can be calculated from Probit equation by:

    A) $Pr = a + b \ln(c/t)$

    B) $Pr = a + b \ln(c.t)$

    C) $Pr = a + b \ln(c+t)$

    D) $Pr = a + b \ln(t/c)$

Answer: B

60. Under which regulation the local authority is responsible for preparation of an offsite Emergency plan?

    A) MSIHC Rules

    B) Hazardous Chemical handling Rules

    C) PMP Acts & Rules

Answer: A

61. Which of the following crisis groups are responsible for preparation of offsite response plan at local level?

    A) State Crisis Group

    B) Local Crisis Group

    C) District Crisis Group

Answer: B

62. The purpose of Hazard maps is to identify and display the location of:

    A) Hazard zones

    B) Risk zones

    C) Vulnerability zones

    D) Decontamination zones

Answer: A

63. Immediate danger area following a toxic gas leak is based upon the substance's:

    A) IDLH

    B) STEL

    C) TLV

    D) Ceiling value

Answer: A

64. The maximum injury causing distance following a toxic gas release will be dependent on the material's:

    A) IDLH

    B) STEL

    C) TLV

    D) Ceiling value

Answer: B

65. Evacuation maps are made based on concentration contours of a toxic release in:

    A) Downwind direction

    B) Cross-wind direction

    C) both the directions

Answer: C

66. In Uncertainty analysis, if the uncertainties are related to the chances of an incident happening, it is called

    A) Aleatory uncertainty

    B) Epistemic uncertainty

Answer: A

67. An area surrounding an incident site where public may be exposed to dangerous and life threatening concentrations of toxic materials is known as:

    A) Initial isolation zone

    B) Protective zone

    C) Decontamination zone

Answer: A

68. An emergency action code designed to help the fire fighters to know readily the actions to be taken in case of fire or spillage of a particular hazardous substance is called:

    A) HAZCHEM code

    B) UN Number

    C) CAS Number

Answer: A

69. How many Hazard Class Labels exist for different hazardous substances?

    A) 10

    B) 9

    C) 7

    D) 5

Answer: B

70. The written emergency information carries by the driver of all vehicles carrying hazardous substances is known as:

    A) HAZCHEM

    B) TREMCARD

    C) Emergency Information pane

Answer: B

71. Emergency Information Panel contains:

    A) Class Lebel

    B) UN Number

    C) HAZCHEM code

    D) All the above

Answer: D

72. The vehicle equipped with rescue and personal protective equipment to be used for rescue purposes is known as:

    A) Rescue vehicle

    B) Rescue trolley vehicle

    C) Rescue Tender

Answer: A or C.

73. The Plan prepared for facilitating the prompt return to the normalcy and continuity of operation after a disaster is called:

    A) Rescue Plan

    B) Relief Plan

    C) Recovery Plan

Answer: C

74. Which of the following is not the part of Disaster Mitigation Planning?

    A) Emergency Communication

    B) Disaster warning system

    C) Risk and vulnerability analysis

    D) Insurance requirements

Answer: A

75. The information required for the preparation of a Disaster Plan is called:

    A) Strategic information

    B) Tactical Information

    C) Operational Information

    D) Statutory information

Answer: A

76. One of the mostly used ways to communicate risk of any hazardous material is:

    A MSDS

    B) Pareto chart

    C) Emergency Information panel

Answer: A

77. Area where head count are usually taken after the evacuation of the affected area is known as:

    A) Shelter area

    B) Head-count area

    C) Mustering area

    D) Rescued area

Answer: C

78. Which of the following is an example of fixed hazard?

    A) Toxic Gas release

    B) Explosion

    C) Pool fire

Answer: C

79. The technique utilized in analyzing disaster response preparedness in industry is called:

    A) HAZOP

    B) ER-HAZOP

    C) What-if

    D) Disaster Mapping

Answer: B

80. The type of personal protection required when the greatest potential for exposure to hazards exists is:

    A) Level-A

    B) Level-B

    C) Level-C

    D) Level-D

Answer: A

81. The first responders who usually experience strong emotional feelings while handling emergencies is said to be under:

    A) Critical stress

    B) Critical Incident Stress

    C) Critical Incident Recall

Answer: B

82. The best way to audit the efficacy of the disaster preparedness is:

   A) Walk-through survey

   B) Table-top exercises

   C) Mock drills at field

Answer: C

83. Which of the following situation arises to the human body while responding to emergencies?

   A) Body rapidly produces Adrenaline

   B) Heart beat increases

   C) Temperature increases

   D) All of the above

Answer: D

Q.1) Dam failures are always catastrophic, especially on the downstream side, resulting in a colossal loss of life and property. Analyse the various causes of dam failures. Give two examples of large dam failures. [150 Words] [10 Marks] **[2023]UPSC**

Q.2) Explain the mechanism and occurrence of cloudburst in the context of the Indian subcontinent. Discuss two recent examples. [150 Words] [10 Marks] **[2022]**

Q.3) Explain the causes and effects of coastal erosion in India. What are the available coastal management techniques for combating the hazard? [250 words] [15 Marks] **[2022]**

Q.4) Describe the various causes and the effects of landslides. Mention components of the important components of National Landslide Risk Management strategy. [250 Words] [15 Marks] **[2021]**

Q.5) Discuss about the vulnerability of India to earthquake related hazards. Give examples including the salient features of major disasters caused by earthquakes in different parts of India during the last three decades. [150 Words] [10 Marks] **[2021]**

Q.6) Discuss the recent measures initiated in disaster management by the Government of India departing from the earlier reactive approach. [250 Words] [15 Marks] **[2020]**

Q.7) Vulnerability is an essential element for defining disaster impacts and its threat to people. How and in what ways can vulnerability to disasters be characterized? Discuss different types of vulnerability with reference to disasters. [150 Words] [10 Marks] **[2019]**

Q.8) Disaster preparedness is the first step in any disaster management process. Explain how hazard zonation mapping will help disaster mitigation in the case of landslides. [250 Words] [15 marks] **[2019]**

Q.9) Describe various measures taken in India for Disaster Risk Reduction (DRR) before and after signing 'Sendai Framework for DRR (2015-2030)'. How is this framework different from 'Hyogo Framework for Action, 2005'? [250 Words] [15 marks] **[2018]**

Q.10) On December 2004, tsunami brought havoc on 14 countries including India. Discuss the factors responsible for occurrence of Tsunami and its effects on life and economy. In the light of guidelines of NDMA (2010) describe the mechanisms for preparedness to reduce the risk during such events. [250 Words] [15 marks] **[2017]**

Q.11) The frequency of urban floods due to high intensity rainfall is increasing over the years. Discussing the reasons for urban floods, highlight the mechanisms for preparedness to reduce the risk during such events. [200 Words] [12.5 Marks] **[2016]**

Q.12) With reference to National Disaster Management Authority (NDMA) guidelines, discuss the measures to be adopted to mitigate the impact of the recent incidents of cloudbursts in many places of Uttarakhand. [200 Words] [12.5 Marks] **[2016]**

Q.13) The frequency of earthquakes appears to have increased in the Indian subcontinent. However, India's preparedness for mitigating their impact has significant gaps. Discuss various aspects. [200 Words] [12.5 Marks] **[2015]**

Q.14) Drought has been recognized as a disaster in view of its spatial expanse, temporal duration, slow onset and lasting effects on vulnerable sections. With a focus on the September 2010 guidelines from the National Disaster Management Authority (NDMA), discuss the mechanisms for preparedness to deal with likely El Nino and La Nina fall outs in India. [200 Words] [12.5 Marks] **[2014]**

Q.15) How important are vulnerability and risk assessment for pre-disaster management? As an administrator, what are key areas that you would focus on in a Disaster Management System? [200 Words] [10 Marks] **[2013]**

14.1 Questions about Tsunamis

1. What is a tsunami?
2. How do earthquakes generate tsunamis?
3. How do volcanic eruptions generate tsunamis?
4. How do submarine landslides, rock falls and underwater slumps generate tsunamis?
5. Can asteroids, meteorites or man-made explosions cause tsunamis?
6. Where and how frequently are tsunamis generated?
7. How does tsunami energy travel across the ocean and how far can tsunamis waves reach?
8. Why aren't tsunamis seen at sea or from the air?

9. What are the factors of destruction from tsunamis?
10. What determines how destructive a tsunami will be near the origin and at a distant shore?
11. What are some of the largest historical tsunamis?
12. The April 1, 1946 Aleutian Earthquake and Tsunami
13. The November 4, 1952 Kamchatka Earthquake and Tsunami
14. The March 9, 1957 Aleutian Earthquake and Tsunami
15. The May 22, 1960 Chilean Earthquake and Tsunami
16. The March 27-28, 1964 Alaska Earthquake and Tsunami
17. Why are locally generated tsunamis so dangerous?
18. Where can I find more tsunami related FAQs?
19. What is a mega-tsunami and can it happen today?

1. What is a tsunami?

The phenomenon we call tsunami is a series of large waves of extremely long wavelength and period usually generated by a violent, impulsive undersea disturbance or activity near the coast or in the ocean. When a sudden displacement of a large volume of water occurs, or if the sea floor is suddenly raised or dropped by an earthquake, big tsunami waves can be formed by forces of gravity. The waves travel out of the area of origin and can be extremely dangerous and damaging when they reach the shore. The word tsunami (pronounced tsoo-nah'-mee) is composed of the Japanese words "tsu" (which means harbor) and "nami" (which means "wave"). Often the term, "seismic or tidal sea wave" is used to describe the same phenomenon, however the terms are misleading, because tsunami waves can be generated by other, non seismic disturbances such as volcanic eruptions or underwater landslides, and have physical characteristics different of tidal waves. The tsunami waves are completely unrelated to the astronomical tides ⊠ which are caused by the extraterrestrial, gravitational influences of the moon, sun, and the planets. Thus, the Japanese word "tsunami", meaning "harbor wave" is the correct, official and all-inclusive term. It has been internationally adopted because it covers all forms of impulsive wave generation.

2. **How do earthquakes generate tsunamis?**

By far, the most destructive tsunamis are generated from large, shallow earthquakes with an epicenter or fault line near or on the ocean floor. These usually occur in regions of the earth characterized by tectonic subduction along tectonic plate boundaries. The high seismicity of such regions is caused by the collision of tectonic plates. When these plates move past each other, they cause large earthquakes, which tilt, offset, or displace large areas of the ocean floor from a few kilometers to as much as a 1,000 km or more. The sudden vertical displacements over such large areas, disturb the ocean's surface, displace water, and generate destructive tsunami waves. The waves can travel great distances from the source region, spreading destruction along their path. For example, the Great 1960 Chilean tsunami was generated by a magnitude 8.3 earthquake that had a rupture zone of over 1,000 km. Its waves were destructive not only in Chile, but also as far away as Hawaii, Japan and elsewhere in the Pacific. It should be noted that not all earthquakes generate tsunamis. Usually, it takes an earthquake with a Richter magnitude exceeding 7.5 to produce a destructive tsunami.

Back to Top

3. **How do volcanic eruptions generate tsunamis?**

Although relatively infrequent, violent volcanic eruptions represent also impulsive disturbances, which can displace a great volume of water and generate extremely destructive tsunami waves in the immediate source area. According to this mechanism, waves may be generated by the sudden displacement of water caused by a volcanic explosion, by a volcano's slope failure, or more likely by a phreatomagmatic explosion and collapse/engulfment of the volcanic magmatic chambers. One of the largest and most destructive tsunamis ever recorded was generated in August 26, 1883 after the explosion and collapse of the volcano of Krakatoa (Krakatau), in Indonesia. This explosion generated waves that reached 135 feet, destroyed coastal towns and villages along the Sunda Strait in both the islands of Java and Sumatra, killing 36, 417 people. It is also believed that the destruction of the Minoan civilization in Greece was caused in 1490 B.C. by the explosion/collapse of the volcano of Santorin in the Aegean Sea.

4. How do submarine landslides, rock falls and underwater slumps generate tsunamis?

Less frequently, tsunami waves can be generated from displacements of water resulting from rock falls, icefalls and sudden submarine landslides or slumps. Such events may be caused impulsively from the instability and sudden failure of submarine slopes, which are sometimes triggered by the ground motions of a strong earthquake. For example in the 1980's, earth moving and construction work of an airport runway along the coast of Southern France, triggered an underwater landslide, which generated destructive tsunami waves in the harbor of Thebes. Major earthquakes are suspected to cause many underwater landslides, which may contribute significantly to tsunami generation. For example, many scientists believe that the 1998 tsunami, which killed thousands of people and destroyed coastal villages along the northern coast of Papua-New Guinea, was generated by a large underwater slump of sediments, triggered by an earthquake.

In general, the energy of tsunami waves generated from landslides or rock falls is rapidly dissipated as they travel away from the source and across the ocean, or within an enclosed or semi-enclosed body of water ⊠ such as a lake or a fjord. However, it should be noted, that the largest tsunami wave ever observed anywhere in the world was caused by a rock fall in Lituya Bay, Alaska on July 9, 1958. Triggered by an earthquake along the Fairweather fault, an approximately 40 million cubic meter rock fall at the head of the bay generated a wave, which reached the incredible height of

520-meter wave (1,720 feet) on the opposite side of the inlet. A initial huge solitary wave of about 180 meters (600 feet) raced at about 160 kilometers per hour (100 mph) within the bay debarking trees along its path. However, the tsunami's energy and height diminished rapidly away from the source area and, once in the open ocean, it was hardly recorded by tide gauge stations.

5. **Can asteroids, meteorites or man-made explosions cause tsunamis?**

Fortunately, for mankind, it is indeed very rare for a meteorite or an asteroid to reach the earth. No asteroid has fallen on the earth within recorded history. Most meteorites burn as they reach the earth's atmosphere. However, large meteorites have hit the earth's surface in the distant past. This is indicated by large craters, which have been found in different parts of the earth. Also, it is possible that an asteroid may have fallen on the earth in prehistoric times ⊠ the last one some 65 million years ago during the Cretaceous period. Since evidence of the fall of meteorites and asteroids on earth exists, we must conclude that they have fallen also in the oceans and seas of the earth, particularly since four fifths of our planet is covered by water.

The fall of meteorites or asteroids in the earth's oceans has the potential of generating tsunamis of cataclysmic proportions. Scientists studying this possibility have concluded that the impact of moderately large asteroid, 5-6 km in diameter, in the middle of the large ocean basin such as the Atlantic Ocean, would produce a tsunami that would travel all the way to the Appalachian Mountains in the upper two-thirds of the United States. On both sides of the Atlantic, coastal cities would be washed out by such a tsunami. An asteroid 5-6 kilometers in diameter impacting between the Hawaiian Islands and the West Coast of North America, would produce a tsunami which would wash out the coastal cities on the West coasts of Canada, U.S. and Mexico and would cover most of the inhabited coastal areas of the Hawaiian islands. Conceivably tsunami waves can also be generated from very large nuclear explosions. However, no tsunami of any significance has ever resulted from the testing of nuclear weapons in the past. Furthermore, such testing is presently prohibited by international treaty.

6. Where and how frequently are tsunamis generated?

Tsunamis are disasters that can be generated in all of the world's oceans, inland seas, and in any large body of water. Each region of the world appears to have its own cycle of frequency and pattern in generating tsunamis that range in size from small to the large and highly destructive events. Most tsunamis occur in the Pacific Ocean and its marginal seas. The reason is that the Pacific covers more than one-third of the earth's surface and is surrounded by a series of mountain chains, deep-ocean trenches and island arcs called the "ring of fire" ⊠ where most earthquakes occur (off the coasts of Kamchatka, Japan, the Kuril Islands, Alaska and South America). Many tsunamis have also been generated in the seas which border the Pacific Ocean.

Tsunamis are generated, by shallow earthquakes all around the Pacific, but those from earthquakes in the tropical Pacific tend to be modest in size. While such tsunamis in these areas may be devastating locally, their energy decays rapidly with distance. Usually, they are not destructive a few hundred kilometers away from their sources. That is not the case with tsunamis generated by great earthquakes in the North Pacific or along the Pacific coast of South America. On the average of about half-a-dozen times per century, a tsunami from one of these regions sweeps across the entire Pacific, is reflected from distant shores, and sets the entire ocean in motion for days. For example, the 1960 Chilean tsunami caused death and destruction throughout the Pacific. Hawaii, Samoa, and Easter Island all recorded runups exceeding 4 m; 61 people were killed in Hawaii. In Japan 200 people died. A similar tsunami in 1868 from northern Chile caused extensive damage in the Austral Islands, Hawaii, Samoa and New Zealand. Although not as frequent, destructive tsunamis have been also been generated in the Atlantic and the Indian Oceans, the Mediterranean

Sea and even within smaller bodies of water, like the Sea of Marmara, in Turkey. In 1999, a large earthquake along the North Anatolian Fault zone, generated a local tsunami, which was particularly damaging in the Bay of Izmit. In the last decade alone, destructive tsunamis have occurred in Nicaragua (1992), Indonesia (1992, 1994, 1996), Japan (1993), Philippines (1994), Mexico (1995), Peru (1996, 2001), Papua-New Guinea (1998), Turkey (1999), and Vanuatu (1999).

7. **How does tsunami energy travel across the ocean and how far can tsunamis waves reach?**

Once a tsunami has been generated, its energy is distributed throughout the water column, regardless of the ocean's depth. A tsunami is made up of a series of very long waves. The waves will travel outward on the surface of the ocean in all directions away from the source area, much like the ripples caused by throwing a rock into a pond. The wavelength of the tsunami waves and their period will depend on the generating mechanism and the dimensions of the source event. If the tsunami is generated from a large earthquake over a large area, its initial wavelength and period will be greater. If the tsunami is caused by a local landslide, both its initial wavelength and period will be shorter. The period of the tsunami waves may range from 5 to 90 minutes. The wave crests of a tsunami can be a thousand km long, and from a few to a hundred kilometers or more apart as they travel across the ocean. On the open ocean, the wavelength of a tsunami may be as much as two hundred kilometers, many times greater than the ocean depth, which is on the order of a few kilometers. In the deep ocean, the height of the tsunami from trough to crest may be only a few centimeters to a meter or more ⊠ again depending on the generating source.

Tsunami waves in the deep ocean can travel at high speeds for long periods of time for distances of thousands of kilometers and lose very little energy in the process. The deeper the water, the greater the speed of tsunami waves will be. For example, at the deepest ocean depths the tsunami wave speed will be as much as 800 km/hr, about the same as that of a jet aircraft. Since the average depth of the Pacific ocean is 4000 m (14,000 feet), tsunami wave speed will average about 200 m/s or over 700 km/hr (500 mph). At such high speeds, a tsunami generated in Aleutian Islands may reach Hawaii in less than four and a half hours. In 1960, great tsunami waves generated in Chile reached Japan, more than 16,800 km away in less than 24 hours, killing hundreds of people.

8. **Why aren't tsunamis seen at sea or from the air?**

In the deep ocean, tsunami wave amplitude is usually less than 1 m (3.3 feet). The crests of tsunami waves may be more than a hundred kilometers or more away from each other. Therefore, passengers on boats at sea, far away from shore where the water is deep, will not feel nor see the tsunami waves as they pass by underneath at high speeds. The tsunami may be perceived as nothing more than a gentle rise and fall of the sea surface. The Great Sanriku tsunami, which struck Honshu, Japan, on June 15, 1896, was completely undetected by fishermen twenty miles out to sea. The deep-water height of this tsunami was only about 40 centimeters when it passed them and yet, when it arrived on the shore, it had transformed into huge waves that killed 28,000 people, destroyed the port of Sanriku and villages along 275 km of coastline. For the same reason of low amplitude and very long periods in the deep ocean, tsunami waves cannot be seen nor detected from the air. From the sky, tsunami waves cannot be distinguished from ordinary ocean waves.

9. **What are the factors of destruction from tsunamis?**

There are three: inundation, wave impact on structures, and erosion. Strong, tsunami-induced currents lead to the erosion of foundations and the collapse of bridges and seawalls. Flotation and drag forces move houses and overturn railroad cars. Considerable damage is caused by the resultant floating debris, including boats and cars that become dangerous projectiles that may crash into buildings, break power lines, and may start fires. Fires from damaged ships in ports or from ruptured coastal oil storage tanks and refinery facilities, can cause damage greater than that inflicted directly by the tsunami. Of increasing concern is the potential effect of tsunami draw down, when receding waters uncover cooling water intakes of nuclear power plants.

10. What determines how destructive a tsunami will be near the origin and at a distant shore?

Tsunamis arrive at a coastline as a series of successive crests (high water levels) and troughs (low water levels) usually occurring 10 to 45 minutes apart. As they enter the shallow waters of coastlines, bays, or harbors, their speed decreases to about 50-60 km/hr. For example, in 15 m of water the speed of a tsunami will be only 45 km/hr. However 100 or more kilometers away, another tsunami wave travels in deep water towards the same shore at a much greater speed, and still behind it there is another wave, traveling at even greater speed. As the tsunami waves become compressed near the coast, the wavelength is shortened and the wave energy is directed upward ⊠ thus increasing their heights considerably. Just as with ordinary surf, the energy of the tsunami waves must be contained in a smaller volume of water, so the waves grow in height. Even though the wavelength shortens near the coast, a tsunami will typically have a wavelength in excess of ten kilometers when it comes ashore. Depending on the water depth and the coastal configuration, the waves may undergo extensive refraction another process that may converge their energy to particular areas on the shore and thus increase the heights even more. Even if a tsunami wave may have been 1 meter of less in the deep ocean, it may grow into a huge 30-35 meter wave when it sweeps over the shore. Thus, tsunami waves may smash into the shore like a wall of water or move in as a fast moving flood or tide ⊠ carrying everything on their path. Either way, the waves become a significant threat to life and property. If the tsunami waves arrive at high tide, or if there are concurrent storm waves in the area, the effects will be cumulative and the inundation and destruction even greater. The historic record shows that there have been many tsunamis that have struck the shores with devastating force, sometimes reaching heights of more than 30-50 meters. For example, the 1946 tsunami generated by an earthquake off Unimak Island in Alaska's Aleutian Islands, reached heights of more than 35 meters, which destroyed a reinforced concrete lighthouse and killed its occupants.

Finally, the maximum height a tsunami reaches on shore is called the runup. It is the vertical distance between the maximum height reached by the water on shore and the mean sea level surface. Any tsunami runup over a meter is dangerous. The flooding by individual waves will typically last from ten minutes to a half-hour, so the danger period can last for hours. Tsunami runup at the point of impact will depend on how the energy is focused, the travel path of the tsunami waves, the coastal configuration, and the offshore topography. Small islands with steep slopes usually experience little runup ⊠ wave heights there are only slightly greater than on the open ocean. This is the reason that islands with steep-sided fringing or barrier reefs are only at moderate risk from tsunamis. However, this is not the case for islands such as the Hawaiian or the Marquesas. Both of these island chains do not have extensive barrier reefs and have broad bays exposed to the open ocean. For example, Hilo Bay at the island of Hawaii and Tahauku Bay at Hiva Oa in the Marquesas are especially vulnerable. The 1946 Aleutian tsunami resulted in runup, which exceeded 8 m at Hilo and 10 m at Tahauku; 59 people were killed in Hilo and two in Tahauku. Similarly, any gap in a reef puts the adjacent shoreline at risk. The local tsunami from the Suva earthquake of 1953 did little damage because of Fiji's extensive offshore reefs. However, two villages on the island of Viti Levu, located on opposite gaps in the reef, were extensively damaged and five people were drowned.

11. What are some of the largest historical tsunamis?

Destructive tsunamis have occurred in all of the world's oceans and seas. In the last half of the 20th Century, Pacific-wide, destructive tsunamis occurred in 1946, 1952, 1957, 1960, and 1964. (Many more tsunamis in inland seas around the periphery of the Pacific, where extremely destructive locally and claimed thousands of lives. Such localized tsunamis occurred in 1975, 1983, 1985, 1992, 1993, 1995, 1998, 1999 and 2001.

The 1 April 1946 Aleutian Earthquake and Tsunami

One of the most destructive Pacific-wide tsunamis was generated by a magnitude 7.8 earthquake near Unimak Island in Alaska's Aleutian Island Chain. A huge wave of 35 meters destroyed completely the U.S. Coast Guard's Scotch Cap lighthouse on Unimak and killed all five of its occupants. The lighthouse was a steel-reinforced concrete structure standing about 30 meters above sea level. Without warning, destructive tsunami waves reached the Hawaiian Islands, five hours later, causing considerable damage and loss of life. The waves completely obliterated Hilo's waterfront on the island of Hawaii, killing 159 people there. Altogether a total 165 people lost their lives from this tsunami, including children attending school at Hawaii's Laupahoehoe Point, where waves reaching up to 8 m destroyed also a hospital. Damage was estimated at $26 million (in 1946 dollars). In 1948, and as a result of this tsunami, the U.S. established a Pacific Tsunami Warning Center in Hawaii.

The 4 November 1952 Kamchatka Earthquake and Tsunami

A strong earthquake (magnitude 8.2) off the coast of Kamchatka Peninsula generated a great destructive Pacific-wide tsunami. Its waves struck the Kamchatka Peninsula, the Kuril Islands and other areas of Russia's Far East, causing considerable damage and loss of life. The tsunami was widely observed and recorded in Japan, but there was no loss of life or damage there. There was considerable damage in the Hawaiian Islands and some damage in Peru and Chile. The tsunami was recorded or observed throughout the islands of the Pacific. In New Zealand waves reached height of 1m. In Alaska, in the Aleutian Islands and in California waves of up to 1.4 meters were observed or recorded.

By far the largest waves outside the generating area were observed in the Hawaiian Islands. Fortunately, no human lives were lost in Hawaii from this tsunami, but damage was extensive, estimates ranging from $800,000-$1,000,000 (in 1952 dollars). The tsunami caused damage on Midway Island. Elsewhere in the Hawaiian island chain, the waves destroyed boats and piers, knocked down telephone lines, and caused extensive beach erosion. In some locations, tsunami waves were destructive in certain locations but hardly noticeable at others. The north shore of the Island of Oahu experienced higher waves of up to 4.5 meters. On the south shore of the island, the tsunami was powerful enough to throw a cement barge in the Honolulu Harbor into a freighter. The island of Hawaii experienced run up to 6.1 meters. In Hilo, a small bridge connecting Coconut Island to the shore was destroyed by one of the tsunami waves lifting it off its foundation, then smashing it down. The effects of the tsunami in the generating area in Kamchatka, varied significantly. From Kamchatka Peninsula to Kronotsky Peninsula the wave heights ranged from zero to 5 meters. From Kronotsky Peninsula to Cape Shipursky the heights ranged from 4-13 meters. The highest wave of 13 meters was the third and was observed at Olga Bay, where it caused considerable damage. Travel time of the first tsunami wave to Olga Bay was approximately 42 minutes after the earthquake. From Cape Shipursky to Cape Povorotny, the tsunami waves ranged from 1 to 10 meters and caused considerable loss of life and damage. At Avachinskaia Bay the tsunami height was 1.2 meters and its travel time was about 30 minutes. From Cape Povorotny to Cape Lopatkka the waves ranged from 5 to 15 meters. At Khodutka Bay a cutter was thrown 500 meters back from shore. On the West coast of Kamchatka Peninsula, the maximum tsunami runup at Ozernoe was 5 meters. At Alaid Island of the Kuril Island group, run up was 1.5 meters. At Shumshu Island it ranged from 7-9 meters. At Paramushir Island the waves ranged from 4-18.4 meters. At Severo ⊠ Kurilsk on Paramushir Island, the second wave was the highest reaching maximum run up of 15 meters. It destroyed most of the town and caused considerable loss of life. At Onekotan Island tsunami run up was 9 meters, while at Shiashkoton Island it was 8 meters and at Iturup Island 2.5 meters. Waves of up to 2 meters were observed at the Komandorsk Islands and at Okhotsk. At Sakhalin ⊠ Korsakov a 1-meter tsunami wave was observed.

The 9 March 1957 Aleutian Earthquake and Tsunami

On March 9, 1957, an 8.3 magnitude earthquake south of the Andreanof Islands, in the Aleutian Islands of Alaska ⊠ in the same general area as that of April 1, 1946 ⊠ generated a Pacific-wide tsunami. Although no lives were lost, there was extensive destruction of property in the Hawaiian Islands, with damage estimated at approximately $5 million (1957 dollars). The waves were particularly high on the north shore of the island of Kauai where they reached a maximum height of 16 meters, flooding the highway and destroying houses and bridges. This was twice the height of the 1946 tsunami. At Hilo, Hawaii, the tsunami runup reached 3.9 m and there was damage to numerous buildings along the waterfront. Within Hilo Bay, Coconut Island was covered by 1 m of water and the bridge connecting it to the shore, as in 1952, was again destroyed.

The 22 May 1960 Chilean Earthquake and Tsunami

The largest earthquake (magnitude 9.5) of the 20th century occurred on May 22, 1960 off the coast of south central Chile. It generated a Pacific-wide tsunami, which was destructive locally in Chile and throughout the Pacific Ocean. The tsunami killed an estimated 2,300 people in Chile. There was tremendous loss of life and property in the Hawaiian Islands, in Japan and elsewhere in the Pacific. Destructive waves in Hilo, Hawaii, destroyed the waterfront and killed 61 people. Total damage was estimated at more than $500 million (1960 dollars).

The 27-28 March 1964 Alaska Earthquake and Tsunami

The largest earthquake of the 20th Century in the northern hemisphere, with a magnitude 8.4, affected an area in Alaska that was almost 1600 km long and more than 300 km wide ⊠ extending from Valdez to the Trinity

Islands, southwest of Kodiak Island in the Gulf of Alaska. The earthquake caused areas to be lifted by as much as 15 m (50 feet) in certain areas, while many other areas subsided greatly. In addition to many local tsunamis generated within the Prince William Sound, vertical crustal displacements averaging 1.8 m (6 ft.) over an area of about 300,000 square kilometers (115,000 square miles) extending in the Gulf of Alaska's continental shelf, generated a Pacific-wide tsunami. Its waves were very destructive in southeastern Alaska, in Vancouver Island (British Columbia), and in the U.S. States of Washington, California and Hawaii. The tsunami killed more than 120 people and caused more than $106 million in damages, making it the costliest ever to strike the Western United States and Canada. Five of Alaska's seven largest communities were devastated by the combination of earthquake and tsunami wave damage. Alaska's fishing industry and most seaport facilities were virtually destroyed. Tsunami waves at Kodiak Island washed away a total of 158 houses and buildings within two blocks of the waterfront. Fishing boats were carried hundreds of meters inland. The 1964 tsunami waves caused also extensive damage in Vancouver Island (British Columbia), and in the states of Washington, California and Hawaii, in the United States. The waves affected the entire California coastline, but were particularly high from Crescent City to Monterey ranging from 2.1 - 6.3 meters (7-21 feet). Hardest hit was Crescent City, California, where waves reaching as much as 6 meters (20-21 feet) destroyed half of the waterfront business district. Eleven persons lost their lives there. At Santa Cruz Harbor, the tsunami waves reached as high as 3.3 meters (11 feet) causing some damage. There was extensive damage in San Francisco Bay, the marinas in Marin County and at the Noyo, Los Angeles and Long Beach harbors. Estimated losses in California were between $1,500,000 and $2,375,000 (1964 dollars), while at Crescent City tsunami damage was estimated at $7,414,000.

12. Why are locally generated tsunamis so dangerous?

A locally generated tsunami may reach a nearby shore in less than ten minutes. There is not sufficient time for the Pacific Tsunami Warning Center or for local authorities to issue a warning. For people living near the coast, the shaking of the ground is a warning that a tsunami may be imminent. For tsunamis from more distant sources, however, accurate warnings of when a tsunami might arrive are possible because tsunamis travel at a known speed

15.1 Question & Answers related to floods

1: What is a flash flood?

A: What is this event actually? How is it different from a flash event? Flash water surges are the most dangerous type of flood because they combine the destructive power of a huge, moving body of water with immense speed. Flash events usually develop rapidly, giving residents little time to evacuate. In particularly steep, clay-earth areas or in regions that receive a lot of rainfall, a flash flood can form in under an hour. Such an event can put countless human lives in danger extremely rapidly.

A "regular" flood, on the other hand, refers to any buildup of water that overcomes the banks, shoreline, levees, beach, embankment, break wall, or other natural or manmade artifacts or formations that generally keep that water at bay. Such a water buildup is still dangerous, as it can submerge homes, cars, people, and animals. But such a disaster event does not usually involve water that is moving as rapidly as in the case of a flash event.

2: What is flood?

A: Rushing water, moving with inertia and a force unparalleled by any machinations of man, it's difficult to comprehend the sheer power of a disaster like this. Floods are the natural disaster that can literally force homes off of their foundations and push them across the ground. Such events can cause immense harm, destroying anything and everything in their path. Of all the natural disasters, these ones also exert the greatest cost in human life. The two worst natural disasters in recorded history were both floods.

The National Severe Storms Laboratory defines events like these as such: "Flooding is an overflowing of water onto land that is normally dry. Floods can happen during heavy rains, when ocean waves come on shore, when snow melts quickly, or when dams or levees break. Damaging flooding may happen with only a few inches of water, or it may cover a house to the rooftop. Floods can occur within minutes or over a long period, and may last days, weeks, or longer. Floods are the most common and widespread of all weather-related natural disasters."

3: Is my house in a flood zone?

A: The best way to determine if your home is in an at-risk zone is to check with your local county or township. They will have maps that indicate which homes are at risk for water surges and which aren't. Keep in mind though, you can still be at risk for water surge even if you don't live in a flood zone.

4: What does flood insurance cover?

A: According to Flood Smart, a program organized by FEMA, "Flood insurance covers losses directly caused by flooding.... Property outside of an insured building. For example, landscaping, wells, septic systems, decks and patios, fences, seawalls, hot tubs, and swimming pools. Financial losses caused by business interruption." It's important to prepare for and protect against such events. Holding an insurance policy that can cover the repairs to your home should not be your first or only line of defense against such disaster events (though it is helpful).

5: What does Flood Zone AE mean?

A: The various types of water risk zones are given specific designations to differentiate them. These designations also determine whether or not homeowners will need to purchase insurance as a part of their mortgage. According to one resource, "AE flood zones are areas that present a 26% chance over the life of a 30-year mortgage, according to FEMA. Since these areas are prone to flooding, homeowners with mortgages from federally regulated lenders are required to purchase flood insurance through the NFIP."

6: What does flood zone x mean?

A: Flood Zone X refers to an area that is outside the 500-year risk area, and which is protected by a levee from the 100-year risk. This is a moderate risk area, meaning that purchase of a home in this area may require insurance for many home mortgage lenders to agree to offer a mortgage for the home.

7: What is Flood Zone A?

A: According to one resource, this region is defined as follows: "Zone A is the flood insurance rate zone that corresponds to the 1-percent annual chance floodplains that are determined in the Flood Insurance Study by approximate methods of analysis. Because detailed hydraulic analyses are not performed for such areas, no Base Flood Elevations or depths are shown within this zone. Mandatory flood insurance purchase requirements apply."

8: When was the flood of Noah?

A: The great water surge event of Noah may have occurred around 2,900 BCE.

9: What is a flood plain?

A: Such a plain is defined by an area of low-lying ground that is adjacent to a river. This geographic region will have been formed mainly by river sediments, and even if it is a good distance away from current river systems, such a plain is still subject to flooding.

10: What is an areal flood warning?

A: Such a warning is usually suggestive of low-risk rainfall and puddling or pooling. Such warnings are usually issued when one to two inches of rain are expected.

11: How to prepare for a flood?

A: You might not have much time to prepare for such a disaster event, so it's important to be hasty and prioritize your time. One government resource suggests that people do as much research and preparation during normal, dry times, so that when a massive water event does occur, they are sufficiently prepared.

According to one organization, "Make a plan for your household, including your pets, so that you and your family know what to do, where to go, and what you will need to protect yourselves from flooding. Learn and practice evacuation routes, shelter plans, and flash flood response. Gather supplies, including non-perishable foods, cleaning supplies, and water for several days, in case you must leave immediately or if services are cut off in your area. Keep important documents in a waterproof container. Create password-protected digital copies. Protect your property. Move valuables to higher levels. Declutter drains and gutters. Install check valves. Consider a sump pump with a battery."

12: What is a flash flood warning?

A: People often wonder what the difference is between a "watch" and a "warning." According to www.weather.gov: "A Flash Flood Warning is issued when a flash flood is imminent or occurring. If you are in a flood prone area move immediately to high ground. A Flood Warning is issued when flooding is imminent or occurring. A Flood Watch is issued when conditions are favorable for a specific hazardous weather event to occur. A Flood Advisory is issued when a specific weather event that is forecast to occur may become a nuisance."

13: How do floods form?

A: One of the reasons why water-related disaster events are the most common type of natural disaster is because the conditions which form such events occur quite frequently. Floods occur when heavy rainfall exceeds the ability of the ground to absorb that sheer amount of rain. Virtually any severe rainstorm has the potential of causing a water surge. Such surges also occur when enough water accumulates in streams, rivers, and lakes for the water to overflow the river banks or the edges of the body of water. Another cause of such events is ice jams and snowmelt. A deep snowpack that melts rapidly can cause flooding to occur. When spring rains exaggerate snowmelt, flash water surge can occur.

14: How do wetlands reduce flooding and erosion?

A: Wetlands are immensely important to the local ecosystem and, as it turns out, to protecting human life and property. Wetlands temporarily store water and slowly release it into the environment. Wetlands are a natural sponge, able to soak up storm waters and prevent water runoff and damage to surrounding human habitation.

15: How to prevent flood?

A: It is almost impossible to guarantee the prevention of such a natural disaster. Ditches, fields, levees, dams, rain gardens, and water retention ponds can help. However, communities must ask the question, "To what extent are we willing to alter the surrounding environment and ecosystem?" It might be more safe to simply not build in extremely water-prone areas, and to instead let nearby wetlands protect the region from excessive water flow.

16: What is a 100 year flood?

A: The name of such an event is an effort to simplify a more complex natural event. Basically, labeling such an event "A 100 year event" is an easy way of categorizing a type of water surge that only has a 1% chance of occurring each year, hence the idea that that level of water flow is only supposed to occur once every 100 years.

17: Where do floods occur?

A: One of the reasons why water flow disasters are the most destructive natural disaster in the United States is because they occur in every U.S. state and territory. They are a threat anywhere in the world that experiences rain. Even arid regions can experience such events because a sudden rainfall can overwhelm dry creek beds and flow into local communities.

18: What causes flash floods?

A: Such events are caused by excessive rainfall in a certain area. They can also be caused by the release of water held by an ice jam or some other type of blockage. Such events usually come about during or after slow-moving thunderstorms that deposit heavy rains on a concentrated area. Disasters like these can also result from hurricanes or other tropical storms.

19: How are flood plains formed?

A: Such plains are formed by erosion, i.e., the gradual cutting of a steam or river into its banks. This process occurs over countless years, sometimes thousands of years, to shape a significant region around a river. As the plain is formed, it can also experience water-level rises during particularly bad storms.

20: How to survive a flood?

A: The best way to protect yourself and others when it comes to dangerous rises in water levels is to be aware and informed of their risk and to have an escape route in place.

1. Know if you live in a flood plain. Are you at risk for a flood? Study local maps and determine if you live in such a plain. Most U.S. counties have detailed flood plain maps, and many counties won't even let new homes be constructed in such plains.

2. Be warned. The best chances for surviving a disaster-level water event come from knowing when a water surge is coming. Invest in an

NOAAA radio, and make sure your cell phone emergency alerts are turned on.

3. Have an escape route. You need to know where you will evacuate to if the water levels begin to rise. Where can you drive to that is higher in elevation than where you currently live? How can you drive there without having to drive through lower elevation areas?

4. Learn the terminology. Know the different types of warnings. The National Oceanic and Atmospheric Association monitors water level conditions and sends out alerts if a water level rise seems likely. For example, a flash event "watch" (or a flood watch) means that flash flooding or flooding is possible within the designated watch area. Be alert. A flash event "warning" (or a flood warning), on the other hand, means that flash flooding or flooding has been reported and is imminent. If you hear such warnings for your area, take necessary precautions at once, and get to higher ground!

5. "Turn around, don't drown." According to the NOAAA, most flooding deaths occur when people attempt to drive their cars into or through water-covered roadways. If you are attempting to evacuate an area and you encounter such a road, turn around and find an alternate route. Never attempt to drive on or through a road that has water flowing over it..

Floods are deadly and dangerous. If such a disaster event is coming your way, do everything you can to evacuate safely. If you cannot evacuate, seek higher ground wherever you are. Never try to swim in or drive through such waters.

21: What does flood zone mean?

A: This simply refers to any region which may experience a water level rise that may be a risk to local human habitation.

22: Why are floods in urban areas particularly dangerous?

A: According to one safety management group, "The main cause of flooding in urban areas is poor or lack of drainage. High intensity rainfall overwhelms sewage and drainage within cities and neighborhoods, filling the streets in a dangerous mix of sewage and floodwater. Urban flooding is a slower process that hinders transportation and may damage daily activities but rarely results in deaths."

23: How to stop flooding?

A: There is not much humans can do to proactively prevent disasters like these from occurring. Rather, it's more about what humans don't do. Case in point, when humans pave over natural soil with parking lots, roads, and buildings, they are making it more difficult for rainwater to seep naturally into the soil. Urban development actually makes serious rainfall worse, not better. Preserving wetlands and investing in water-pervious paving methods can prevent serious disaster events from occurring.

24: What does 100 year flood plain mean?

A: This refers to an area that is only expected to receive higher-than-normal water levels once every 100 years. Another way to look at it is that such regions only have a 1% chance of experiencing disaster-level water events each year.

25: What does flash flood mean

A: This is simply a type of water event that involves the rapid rise of water levels and disbursement of water into a region with little to no warning.

26: What does flood plain mean?

A: The term simply refers to an area in and around a river or stream that is subject to rising water levels during excessive rain and other natural events.

27: What is a flood tide?

A: This refers to a powerful surge of water onto land, usually tidal water in the ocean.

28: What is flood zone c?

A: This refers to an area with minimal hazard for serious water runoff events.

29: What to do during a flood?

A: Turn around, don't drown! That is the catch phrase to remember. If safe evacuation over roads that do NOT have water on them is possible, then evacuate. But if roads to higher ground are not passable, turn around, and try to find higher ground somewhere else. Sometimes, the highest ground available will be the second floor or roof of you home, so keep that in mind as a potentially good place to wait out a storm event.

30: How does the formation of a natural levee impact flooding?

A: Natural levees can help prevent excessive water events from occurring. However, if the natural levees fail, it can often result in a serious water runoff event.

31: How long do floods last?

A: According to the Flood Observatory (recording such events since 1985), the average duration of water above normal levels is 9.5 days. While floods can form very quickly (particularly in the case of flash floods), they often linger for many days at a time before the water level returns to a normal range. A flood is still quite dangerous, even if the water is no longer moving as rapidly.

32: What are the effects of flood?

A: Such an event can bring walls of water from ten to twenty feet high. A car can be taken away by such water in as little as two feet of water, and entire homes can be shifted off their foundations by particularly strong surges.

Such a disaster event can cover a large swath of land too, depending on the geography of the region.

33: What is flood stage?

A: This term refers to the water level, as read by a stream gauge or tide gauge, for a body of water at a particular location, at a particular time. As the measurement varies, experts are able to monitor the various stages of excess water flowing into or out of that region at any given time.

34: What is the definition of flood?

A: The National Severe Storms Laboratory defines these events as such: "Flooding is an overflowing of water onto land that is normally dry. Floods can happen during heavy rains, when ocean waves come on shore, when snow melts quickly, or when dams or levees break. Damaging flooding may happen with only a few inches of water, or it may cover a house to the rooftop. Floods can occur within minutes or over a long period, and may last days, weeks, or longer. Floods are the most common and widespread of all weather-related natural disasters."

35: Where do floods happen?

A: Such events are most likely to occur in areas that experience significant rainfall or in coastal and river regions. But disasters like these can technically occur anywhere that has rainfall.

36: What is a 500 year flood?

A: A 500 year event such as this is one that only has a 0.2% chance of occurring each year, hence the likelihood of such a disaster event only occurring once every 500 years.

37: What to do after a flood?

A: Do not enter standing water that is leftover from such an event, as it may be carrying an electrical current. Wait until all of the water has receded before you begin the cleanup and recovery process.

38: What to do if your house floods from rain?

A: If the water surge is so bad that water begins to come into your home, there are several things you must do, and quickly! Turn off the electricity to your home, to prevent electrical currents from traveling through the water. Evacuate the premises if you are able to plot out a safe route along a road that does NOT have water flowing over it. If that is not possible, call for help. While waiting for help to arrive, seek higher ground in the second floor of your home, or on the roof, if necessary.

39: What to do in a flash flood?

A: There are some basic, life-saving rules to follow during a sudden water surge event. From the Centers for Disease Control and Prevention: "Gather emergency supplies, including food and water. Store at least 1 gallon of water per day for each person and each pet. Store at least a 3-day supply. Listen to your local radio or television station for updates. Have immunization records handy (or know the year of your last tetanus shot). Bring in outdoor possessions (lawn furniture, grills, trash cans) or tie them down securely. If evacuation appears necessary, turn off all utilities at the main power switch and close the main gas valve. Leave areas subject to flooding such as low spots, canyons, washes, etc. (Remember: avoid driving through flooded areas and standing water.)"

40: What is coastal flooding?

A: This is an event in which low-lying coastal regions are submerged by sea water. This can occur as a result of hurricanes and other tropical storms, or as a result of particular high tides or storm surges.

41: Where do floods mostly occur?

A: River plains and coastal regions are the areas most susceptible to such disaster events. However, a disaster like this can actually occur just about anywhere that experiences rainfall. Even in the desert, immense water runoff has been known to occur.

42: Why are flash floods so dangerous?

A: They are primarily dangerous because they occur so rapidly and can often catch people off guard. Particularly in rural areas with few roads, such a disaster event can quickly trap people in an isolated region, often giving them no way to safely evacuate.

43: How can paving over a wetland area cause flooding?

A: Wetlands are extremely important! They should not be altered, drained, or paved over. They act as natural sponges that absorb excess rainwater and stormwaters. Paving over them makes water runoff worse, creating more severe water surge events as a result.

44: How deadly is 1931 China floods?

A: According to the records, the 1931 flooding of the Yangtze River in China (called the Central China Floods of 1931) comes first for the worst natural disaster in human history. Anywhere from 2 million to 3.7 million people died because of this flood. One of the reasons for the huge death toll is that the flood covered a massive swath of land, about 70,000 square miles all in all.

45: What makes floods different from high water levels?

A: One of the reasons why water level rises are the most common type of natural disaster is because the conditions which form such events occur quite frequently. Surges occur when heavy rainfall exceeds the ability of the ground to absorb that sheer amount of rain. Virtually any severe rainstorm has the potential of causing such an event. Floods also occur when enough water accumulates in streams, rivers, and lakes for the water to overflow the river banks or the edges of the body of water. Another cause of floods is ice jams and snowmelt. A deep snowpack that melts rapidly can cause surging to occur. When spring rains exaggerate snowmelt, flash flows can occur.

46: What damage can a flood cause?

A: Floods can cause immense damage to anything in their path. Flash floods, in particular, are quite devastating, as they tend to involve fast-moving water. Densely populated areas are also at higher risk for crisis-level incidences because buildings, highways, and parking lots increase runoff by reducing the amount of rain absorbed by the ground.

If a thunderstorm lingers over an area for an extended period, it can cause a nearby stream with just a few inches of water in it to rise ten feet or more in under an hour. That poses a particular risk to people who are camping or recreating near bodies of water during intense rainfall.

The primary cause for concern with floods are the legitimate risk of drowning. Such events occur quickly, they usually involve fast-moving water, and they are quite unpredictable. People who get caught in a disaster like this are often unable to get out of the water, and they drown. In the United States, floods kill more people than tornadoes, hurricanes, or lightning

47: What areas are most at risk from flash floods?

A: All such events are dangerous, but flash floods are by far the most dangerous. With that in mind, the following are the regions of the U.S. where flash floods are most likely to occur:

Densely populated areas. As mentioned earlier, the buildup of concrete structures, essentially "paving over the world," makes stormwater runoff conditions much worse.

Areas near rivers. These regions are at risk for serious water level rises. Even when levees are built near flood-prone rivers, such installations are not always effective at preventing the water level from rising over the levee.

Dam failures. When a dam fails, this causes a sudden and destructive surge of water to plummet downstream. Such an event can be particularly devastating, as people living nearby usually have little to no warning of such an event occurring.

Mountainous regions, steep hills, and clay-like soil. Steep areas are more prone to water level rises, particularly in the southeastern United States, where the soil is more clay-like and not as absorptive.

Canyons and river beds. Out west, a dry creek bed or a canyon may seem like a safe place to recreate. But this is not always the case. During incidences of heavy rainfall (even rainfall occurring far away from campers upstream), dry creek beds and canyons can fill with water rapidly, posing an immense risk to those nearby.

Recent burn areas. An area that has been burned recently (such as a coordinated burning in a forest) is at risk for water surges, as much of the vegetation which would have captured flood waters will have been burned away.

48: Can flood waters resurge and rise again after falling?

A: Yes, they absolutely can! One of the more sinister risks of water level rises is that the water levels can begin to recede, and there can be the appearance of safety. Then, suddenly, a second rainstorm can occur upstream, or a levee can break, or a dam fails, and floodwater levels rise again. This is why it is critical not to approach a recently flooded area until public health and safety experts say it is safe to do so.

49: What are the most dangerous floods?

A: The two most devastating natural disasters in recorded history were both the result of sudden increases in water levels. According to the records, the 1931 flooding of the Yangtze River in China (called the Central China Floods of 1931) comes first for the worst natural disaster in human history. Anywhere from 2 million to 3.7 million people died because of this disaster event. One of the reasons for the huge death toll is that the flood covered a massive swath of land, about 70,000 square miles all in all.

The second-worst natural disaster in human history was also a flood, and it also occurred in China. This was the 1887 Yellow River Flood, a natural disaster that killed anywhere from 900,000 to 2 million people. The event was partially caused by human interference, as the Yellow River had risen above and away from nearby farmland by a series of dikes. When heavy rains surged the river, it spilled over the dikes and put about 5,000 square miles underwater.

As for such disasters in the United States, the worst one was undoubtedly the 1889 flood in Johnstown, Pennsylvania. In this disaster, an upstream dam failed, and a wall of water 40 feet high and a half-mile wide came roaring down upon the Appalachian town of Johnstown, killing 2,209 people within minutes.

50: Are floods getting worse?

Most scientists believe that instances of water surge will become more frequent and more intense in the coming years. As it currently stands, an increasing number of coastal and inland communities are already experiencing higher numbers of events of this kind.

There is a growing body of evidence that, as the effects of climate change become more pronounced, floods will become more frequent and more intense. Quoting the Natural Resources Defense Council, "As the IPCC (Intergovernmental Panel on Climate Change) noted in its special report on extremes, it is increasingly clear that climate change 'has detectably influenced' several of the water-related variables that contribute to floods, such as rainfall and snowmelt. In other words, while our warming world may not induce floods directly, it exacerbates many of the factors that do. According to the Climate Science Special Report (issued as part of the Fourth National Climate Assessment, which reports on climate change in America), more flooding in the United States is occurring in the Mississippi River Valley, Midwest, and Northeast, while U.S. coastal flooding has doubled in a matter of decades."

Sources:

References

1. Abe K. (1995). *Estimate of Tsunami Run-up Heights from Earthquake Magnitudes.* ISBN 978-0-7923-3483-5.
2. Haugen, K; Lovholt, F; Harbitz, C (2005). "Fundamental mechanisms for tsunami generation by submarine mass flows in idealised geometries". *Marine and Petroleum Geology.* 22 (1–2): 209–217. doi:10.1016/j.marpetgeo.2004.10.016
3. Lekkas E.; Andreadakis E.; Kostaki I.; Kapourani E. (2013). "A Proposal for a New Integrated Tsunami Intensity Scale (ITIS-2012)". *Bulletin of the Seismological Society of America.* 103 (2B): 1493–1502. doi:10.1785/0120120099
4. Levin, Boris; Nosov, Mikhail (2009). *Physics of Tsunamis.* Dordrecht: Springer. ISBN 978-1-4020-8855-1.
5. Voit, S.S (1987). "Tsunamis". *Annual Review of Fluid Mechanics.* 19 (1): 217–236. doi:10.1146/annurev.fl.19.010187.001245
6. The science behind tsunamis | National Oceanic and Atmospheric Administration
7. Disaster Management Previous Year Questions (PYQs) | UPSC Mains Examination |ForumIAS
8. Frequently Asked Questions about Tsunamis - TeacherVision
9. https://www.nssl.noaa.gov/education/svrwx101/floods/
10. https://floodobservatory.colorado.edu/archiveatlas/duration.htm#
11. https://www.livescience.com/33316-top-10-deadliest-natural-disasters.html
12. https://www.history.com/news/deadliest-natural-disasters-us-storm-flood-hurricane-fire
13. https://www.floodsmart.gov/coverage

14. https://www.amica.com/en/products/flood-insurance/what-is-an-ae-flood-zone.html#

15. https://www.selective.com/~/media/Files/S/Selective/documents/Flood/Forms/what%20are%20the%20different%20zones.pdf

16. https://www.ready.gov/floods

17. https://www.weather.gov/safety/flood-watch-warninghttps://safetymanagement.eku.edu/blog/flood-basics-and-safety/

18. The Five R's: The Basics of Sustainability - Sustainability - Green Notes

19. National Overview: Facts and Figures on Materials, Wastes and Recycling | US EPA

20. Functions and Types of Ecosystems

21. reycycling techniques notes - Search

22. plastic recycling - Search Images

23. Environmental Science Archives -

www.ingramcontent.com/pod-product-compliance
Ingram Content Group UK Ltd.
Pitfield, Milton Keynes, MK11 3LW, UK
UKHW060404300726
14090UKWH00006B/432

9798899841484